Urban Transformations

Cities affect every person's life, yet across the traditional divides of class, age, gender and political affiliation, armies of people are united in their dislike of the transformations that cities have undergone in recent times. The physical form of the urban environment is not a designer add-on to 'real' social issues; it is a central aspect of the social world. Yet in many people's experience, the cumulative impacts of recent urban development have created widely un-loved urban places. To work towards better-loved urban environments, we need to understand how current problems have arisen and identify practical action to address them.

Urban Transformations examines the crucial issues relating to how cities are formed, how people use these urban environments and how cities can be transformed into better places. Exploring the links between the concrete physicality of the built environment and the complex social, economic, political and cultural processes through which the physical urban form is produced and consumed, Ian Bentley proposes a framework of ideas to provoke and develop current debate and new forms of practice.

The book focuses on four key questions, examining the most helpful conceptual framework for thinking about the processes of urban transformation; how this framework can help us to understand how these processes generate, through speculative markets, the forms and patterns of land use which typify recent urban places; the common ground for change that can be identified as the practical basis for widespread action towards better-loved and more sustainable urban places; and how the necessary changes can actually be made to happen in cities world-wide. Ian Bentley concludes by identifying the most promising types of urban forms and working practices through which users and professionals might work together to develop better-loved urban places in the future.

Ian Bentley is professor of Urban Design at the Joint Centre for Urban Design, Oxford Brookes University.

Urban Transformations

Power, people and urban design

Ian Bentley

London and New York

First published 1999
by Routledge
11 New Fetter Lane, London EC4P 4EE

Simultaneously published in the USA and Canada
by Routledge
29 West 35th Street, New York, NY 10001

Reprinted 2001, 2002

First published in paperback 2002

Routledge is an imprint of the Taylor & Francis Group

© 1999, 2002 Ian Bentley

Typeset in Garamond by J&L Composition Ltd, Filey, North Yorkshire
Printed and bound in Great Britain by
Biddles Ltd, Guildford and King's Lynn

British Library Cataloguing in Publication Data
A catalogue record for this book is available from the British Library

Library of Congress Cataloging in Publication Data
A catalog record for this book has been requested

ISBN 0–415–12823–4 (Hbk)
ISBN 0–415–12824–2 (Pbk)

For Iva

Do not think that one has to be sad in order to be militant, even though the thing one is fighting is abominable. It is the connection of desire to reality (and not its retreat into the forms of representation) that possesses revolutionary force.

Michel Foucault

Contents

Figures

Acknowledgements

It has taken me a very long time to get this book together, and I have incurred a mass of intellectual and emotional debts along the way. It gives me pleasure to acknowledge these now.

Every project has to start somewhere. This one grew from a long-running dialogue with Christina Dorees, who convinced me that I should try to construct a big argument from what had previously been a collection of scattered insights. Her painstaking reviews of early drafts, with the aid of her built-in crap detector, had a key influence on the argument's final form.

Nothing grows out of thin air, however, and I want to acknowledge the major influence and support which I have drawn from earlier work in the Responsive Environments group at Oxford Brookes's Joint Centre for Urban Design. In particular, debates and practice relationships with Alan Alcock, Paul Murrain, Sue McGlynn and Graham Smith were central in creating the original cultural context within which this book developed.

During the progress of the project, many people gave generous help and advice. Georgia Butina-Watson has been a close collaborator throughout: I have drawn freely on insights from a range of seminar series which we and our graduate students developed together. Peter Ambrose, Sue McGlynn and Ed Robbins read the penultimate draft, and between them inspired a major and much-needed shift of emphasis; whilst Roger Simmonds made many detailed and useful comments on Part II.

Writing a book is not just a matter of thinking up ideas. One of the hardest and most demanding aspects is concerned with managing the production of the manuscript as a coherent physical artefact, in the context of anarchic absent-mindedness which seems to permeate my own work. I feel great admiration and gratitude for three remarkable people who supported me when I needed them. Elaine Butler gave me the enthusiastic help to get the project off the launchpad, and Angela Ko has been tireless in her attempts to keep it from crash-landing at the end (and funny with it). Finally, it gives me the greatest pleasure to say thank you to Linda New. She has created a climate of friendship and encouragement far beyond anything I had a right to expect, and *Urban Transformations* would not have happened without her. I am very grateful.

The author and publishers would like to thank the following individuals and organisations for permission to publish the figures noted below:

Figure 1.1, Scott Abramowitz; Figure 1.2, Satish Mahindra; Figure 1.3, Dikran Nalbandian; Figure 1.4, Paola Guidugli; Figure 1.5, Harvard University Press; Figure 1.6, Graham Paul Smith; Figure 2.3, Oxford City Council; Figure 3.2, Penguin Books/Edward de Bono; Figure 3.3, Lionel March and Philip Steadman; Figure 4.6, Angela Ko; Figure 5.1, Sidgwick and Jackson; Figure 6.2, Reinhold Publishing Corporation; Figure 6.3, Butterworth Heinemann; Figure 7.1, Binoy Panicker; Figure 8.1, Helen Teague; Figure 8.2, Timothy Long; Figure 8.3, American Planning Association/Newman and Kenworthy; Figure 9.2, Christina Dorees; Figure 9.3, Lora Nicolaou; Figure 9.8, Butterworth Heinemann, from *The Concise Townscape* by Gordon Cullen; Figure 9.9, Architects' Journal/Bill Hillier; Figure 9.11, Georgia Butina Watson; Figure 10.4, Amsterdam City Council; Figure 10.6, David Watson; Figure 10.7, Amsterdam City Council; Figure 10.8, Graham Paul Smith; Figure 10.10, Graham Paul Smith; Figure 10.13, Graham Paul Smith; Figure 10.15, Journal of Architectural and Planning Research; Figure 10.16, Graham Paul Smith; Figure 10.17, Jannes Linders; Figure 10.18, Stirling Wilford and Partners; Figure 10.19, Architects' Journal; Figure 10.20, Mills Beaumont Leavey Channon; Figure 10.21, Alan Reeve.

All other figures are by the author.

Introduction

The best moments of my life have mostly been in cities. It is in cities, by and large, that I have felt my spine tingle at the sheer beauty of places, and at the variety of experiences they can offer. And it is only in cities that I have experienced those moments of being alone in a friendly crowd that offer the sense of being at once an individual and a member of the human race. As you can tell, I am a romantic so far as cities are concerned. This book is the offspring of an urban love affair.

As someone who loves cities, I find myself alternatively exhilarated, saddened and angered by their recent transformations. On the one hand, for example, I am exhilarated by the increase in multi-cultural richness which I have experienced during my own lifetime. On the other hand, I am saddened and angered by other aspects of urban change, and I am not alone here: Alison Ravetz, for example, surely speaks for many when she says that: 'we are now in a period where most modern development is disparaged and modern architecture frequently described as "a crime against humanity", while the large city is increasingly associated, as it was in Victorian times, with poverty, violence and danger.'[1]

In no way, however, should we drift into reading the process of urban transformation as a doom-and-gloom story. Even its most negative aspects have positive, inspirational outcomes, as people from all walks of life fight back to restore the life-enhancing potentials which urban places can have. The amount of time and effort which many people put into improving unloved places, or protecting loved ones from negative change, bears witness to an active resilience which is not the preserve of any particular social group. Those who are most articulate and better-off have more resources to deploy, but the heroic efforts which are also made by under-resourced community groups working to improve modern social housing estates, for example, show that the willingness to fight for better-loved places is not confined to the chattering and scribbling classes.

Inspirational though these efforts are, however, very few people engage in them for fun or in search of moral uplift. I have been professionally involved in a number of them over the years, and I have consistently found that they are mostly carried out by people who would rather be doing something else, but are forced into action because they feel that there is no alternative if they are to have acceptable urban settings for their lives. All this effort, then, is evidence of some kind of mismatch between the processes through which urban form is produced and transformed, and the desires and aspirations of its users. Speaking with unlikely agreement, across traditional barriers of class, age, political affiliation, ethnicity and gender, city-dwellers are insisting by their actions that something has gone seriously wrong.

This is a situation which cries out for change. It raises serious ethical issues for developers, government agencies and design professionals alike, and it is also a matter of grave economic importance. Too many massive schemes have been demolished because they quickly proved uninhabitable: St Louis's award-winning Pruitt Igoe estate, blown up in 1972, was only the first and most notorious of many on both sides of the Atlantic. Too many others have had to be modified in drastic ways, often at prices far more than their original costs; making it clear that unloved urban places are also awesomely bad investments. From the bleeding-heart liberal to the driest of free-marketeers, we all need to engage in the effort to get better-loved places on the ground.

At the end of the second millennium, this imperative feels ever more urgent. On the one hand, we see the growth of widespread fears about the role which current patterns of urban transformation are playing in an ecological crisis which, some fear, may ultimately threaten the future of human life itself. On the other hand, these dire fears have their positive side, for they foster the development of a powerful 'environmental lobby', with international political support, which offers the prospect of an altogether new scale of opportunity-space for rethinking urban futures.

If we are to take advantage of these new opportunities, we need optimism rather than a paralysing culture of fear. This is no time, however, for the simplistic thinking which often accompanies too facile an optimism. The issues to be thought through here are complex and multi-dimensional. In practice, however, they are all too often addressed in myopic, one-dimensional ways. According to the particular viewpoints of one or other of the myriad specialists in urban affairs, complex issues are reduced to economic problems, or social problems, or architectural problems or whatever. As one particularly blinding aspect of this myopia, urban decisions are typically made with little understanding of the complex interactions between the social and the spatial worlds, ignoring the fact that cities, amongst their other attributes, are the unavoidably spatial settings of our unavoidably social lives, which they affect in far-reaching ways.

Perhaps precisely because it has such an all-pervasive impact on human affairs, the spatial aspect of social experience is all too often just taken for granted in a completely unfocused way. To bring it into focus for a moment, consider your own situation right now. Most of you, no doubt, will be reading these words in a room of some kind. At one level, the design of the room supports the activity of reading, and other activities besides. As I write these words, it is raining. If there were no roof over my head, the paper would dissolve.

This ability to support human activities is a crucial aspect of any built form. After all, places are usually constructed, at least in part, because they facilitate activities which would be difficult, if not impossible, in a wilderness. At another level, however, the design of any place also puts constraints on the activities which take place in and around it. For example, you will almost certainly have entered your room through a doorway. It is extremely unlikely that you will have smashed your way in through a wall, or tunnelled through the floor; cultural rules, legal sanctions and lack of resources would prevent you, except perhaps in times of war. And if you grow bored with these pages, and wish to glance out at a longer view, you will have to look out through a window, rather than through the solid bits of the walls, floor or ceiling. It is clear that the physical form of the room – the relative positions of its solids and voids – affects what its users can and cannot do: despite its silence, a wall says 'no'; and it says it louder than any 'no' you will ever be able to hear. The spatial structure of the room, and of its relationship to the rest of the built world, has a political dimension.

The politics of space also has an ideological dimension, arising because people invest places with meanings related to rules which they have learned through their own cultural backgrounds. These meanings may impose further constraints on the users concerned, or may offer them further opportunities. For example, imagine two rooms which have identical dimensions and lighting levels, heated to the same comfortable temperature. Imagine that one has windows on to a mountain landscape, walls painted with murals, and an inlaid marble floor. Imagine that the other has walls of unpainted cinder block with a bare concrete floor, and that it only has artificial lighting. One space is to be used as a living room, and the other as a garage. Which, in your culture, would be used for which purpose?

This is, of course, a rhetorical question with an unrealistically obvious answer, chosen to make a point. Real-life examples are often more ambiguous, because different users may invest given places with different meanings, related to the ideas which they have developed to make sense of their different life-experiences. The 'landmark' office block, whose image might support the managing director's liberating sense of being 'in command', might equally reinforce the typist's claustrophobic feeling of being reduced to a small cog in the corporate machine. Through its ideological dimension, the place suggests opportunities to the one but mostly constraints to the other.

Ultimately, neither the physical nor the ideological structuring of space can determine what people do: even a prison needs its gaolers to make its inmates into prisoners, and even the Berlin Wall needed its guards. Together, however, these physical and ideological factors do offer opportunities and they do exert constraints. These opportunities and constraints may have only a trivial importance in the case of most individual rooms. But when the rooms are combined to make up whole buildings, and when the buildings are assembled into cities, the cumulative pattern of opportunities and constraints has important social and economic impacts, looming large in many people's lives. To take an extreme example, the Pruitt-Igoes of the world are not demolished because there is no demand for the low-rent housing they contain. They are demolished because they provide low-rent housing in an inappropriate form – a form which, in the social context concerned, offers too few opportunities which people might call on to improve their lives, and which imposes too many constraints on people's efforts to deal with the particular economic and social problems which they face.

The physical form of the urban environment, then, is not some sort of designer add-on to 'real' social issues. It is a central aspect of the social world itself, contributing to the constitution of that world through every dimension from the economic through the biotic to the aesthetic. And, in many people's experience, it has somehow gone wrong through the cumulative impacts of recent urban development. Somehow, we have to find ways of working towards better-loved places.

Whenever we try to decide how the future ought to be, we are essentially dealing in predictions, and we know as a matter of fact that most predictions turn out to be wrong. Because cities affect all our lives in important ways, we need the best predictions we can get. Since urban systems are so complex, good predictions will have to incorporate knowledge and ideas from many different fields. They are much more likely to arise from a wide-ranging debate between people from all sorts of different backgrounds, than from the thoughts of any single individual, no matter how wise. The purpose of this book, therefore, is to propose a framework of ideas to help in provoking a debate around four main questions. First, what is the most helpful *conceptual framework* for thinking about the processes of urban transformation? Second, how does this framework help us understand how these processes generate their particular *product*: the forms and patterns of land use which typify recent urban places? Third, what *common ground for change* can we identify, as the practical basis for widespread action towards better-loved and more sustainable places? Finally, *how can we make these changes happen?* The book's four parts address these questions in turn.

Part I

Problematics of production

Introduction

We cannot make better urban places, except by happy accident, unless we have a reasonable grasp of how built form is produced. As a first step, we must build a framework of ideas which we can use to develop this understanding.

Because of my job, I discuss the production of built form with a wide range of people: community activists and residents of run-down social housing, as well as practitioners and academics from many professional backgrounds. From these discussions, it is quite clear that all these people, from all these different backgrounds, already have their own ideas about how urban form is produced. Rather than offering a fully-fledged explanation, however, it seems to me that each of these conceptions has only the character of a problematic: 'the sense of significance and coherence one brings to the world in general in order to make sense of it in particular'.[1]

None of these problematics could have survived for long, in the mainstream culture of any practical field, unless it had a practical value to the workers there. This has two implications from our point of view.

First, on the positive side, it tells us that all mainstream problematics must have some potential for helping us understand how the form-production process works in practice. Activists, developers and urban change professionals alike spend their working lives in a continuous series of action-research experiments, to discover what they can and cannot achieve in the form-production process; whilst academics have theoretical concepts which can be used to organise these practical insights together. All these problematics have something positive to offer.

On the other hand, from our point of view, this practical utility has a negative potential too. As well as containing indispensable practical knowledge about what can and cannot be done, it must also enable practitioners to feel sufficiently good about what they do, in psychological terms, to be able to go on doing it: in the case of current form-production, that is, to go on being implicated in the production of unloved urban places.

We do not need a raft of theory at this point to see that those of us who are involved in producing these places might not automatically feel good in this situation, and that one obvious coping mechanism might involve the development of problematics which help us not to notice those aspects of what we do which, if we did notice them, might make us feel unable to go on. This implies the distinct possibility that practical problematics might sometimes be constructed to hide as much as they reveal.

The negative sides of practical problematics, however, are as valuable as the positive ones in constructing a useful understanding of how form is produced. A key part of understanding the production process, after all, involves grasping how producers hide from themselves the negative impacts of what they do, so that they can go on producing.

To make use of either the positive or the negative aspect of practical problematics, however, we have to discover which is which: which aspect can help us explain the workings of the form-production process, and which can enlighten us about how producers cope with the contradictions of their working lives by obscuring how the process works. If we muddle these up, of course, we are in dire straits. To yield their useful insights, therefore, practical problematics have to be subjected to a critical review – to see which aspects can be used to explain the empirical evidence of the real world, and which are better understood as mechanisms to help practitioners cope with the contradictions which form-production entails.

There is a wide range of practical problematics in good currency today, associated with the complex division of labour in today's form-production process. Together, these problematics form a sprawling mass of overlapping conceptions, often appearing to contradict one another, but all with some value for us. To review this complex field in a book, we have to find some way of organising it into a relatively simple pattern of chapters, structured around the factors which interest us here – the factors which, according to each problematic, might empower or constrain our intervention in the form-production process.

In deciding the structure of the various chapters, and in organising the arguments within them, I have adopted a straightforward strategy. We shall start with whichever problematic offers the simplest account of the factors which interest us, reviewing this to discover both its strengths and its shortcomings from our point of view. The shortcomings thus revealed will guide us towards other more complex problematics which offer ways of remedying them; and so on until we have gone as far as we can towards a coherent and satisfying explanation of how form-production works in our terms, integrating all the key insights which the earlier reviews have revealed.

The simplest group of problematics, in our terms, contains those which see built form as generated through forces external to human action, playing down particular people's interventions in its production. These problematics are reviewed in Chapter 1. They are strong on the factors which *constrain* intervention in the form-production process, but weak on those which might *empower* it.

To redress this shortcoming, Chapter 2 reviews a range of problematics which conceive of built form as generated through the actions and interactions of particular individuals. This approach too yields many useful insights, but is unable to explain how the form-production process is held together as a co-ordinated whole, generating built form which is strongly patterned, rather than appearing merely as the chaotic result of idiosyncratic actions. The analysis here shows that problematics of this kind have to be supplemented by some explanation of the factors through which this co-ordination works.

Chapter 3 identifies the 'problematic of type' as practical culture's only current set of ideas for understanding this co-ordination process. Here co-ordination is seen as taking place through the medium of cultural structures called 'types'. Each type embodies the minimum, irreducible specification for some element of built form in good currency, and constitutes both the cultural raw material which makes the form-production process possible in the first place, and rules which structure its physical outcomes. This is an altogether broader, more inclusive approach than those explored in the earlier chapters, and we shall see how it can be used to overcome their weaknesses whilst synthesising their positive insights.

By the end of this first part of the book, we shall have developed a framework of ideas for understanding the practical workings of the form-production process, as the first step towards changing it.

1 Untouched by human hand (well, almost)

Many people involved in producing urban places call on problematics which incorporate the idea that built form is shaped by forces which originate outside the human world, imposing themselves on physical form whether we want them to or not. This idea comes in various guises, intersecting and overlapping each other in a complex web within form-production culture as a whole.

The simplest version of this approach, from our point of view, is that of the *Zeitgeist* or 'Spirit of the Age', according to which history itself has 'needs' and follows certain inexorable laws. 'History is destiny, unfolded through the life and death of cultured human societies', as the German historian Oswald Spengler put it.[1] Cultures are born, flower and decay in time with the rhythms of destiny, and built form changes with them.

A second version, less simple from our perspective, sees the production of form as conceived in Mother Nature's womb. Here it is various aspects of the natural world, such as climate and the natural, physical characteristics of building materials, from which built form is born.

In a third guise, more complex still, this problematic takes the view that built form is determined by technology. This idea, celebrated in books with titles like *Mechanisation Takes Command*,[2] is deeply embedded in the problematics which many people call on to make sense of urban form. As Peter Ambrose notes: 'It was "the railways" that came in the middle of the last century and took so much city-centre land for their termini. It is "the motor car" that is generating the demand for out-of-town shopping centres'.[3]

Fourthly, at the most complex level from our point of view, there is the widespread version of this problematic which holds that 'form follows function' – surely the best-known slogan in modern design culture. As the American architect Louis Sullivan put it many years ago, 'it was not simply a matter of form expressing function; the vital idea was that the function created or organised its form.'[4]

I have sketched out these four versions of the 'external forces' problematic as conceptually distinct, but in practice their interactions are more complex than this simple picture makes out. Not only do the various versions overlap and intersect with each other, but they are also often called on in 'weaker' forms, which allow at least some scope for human action.

This complexity is well portrayed by the British architect Neville Conder:

> the modern architect returns to first principles. Without preconceived ideas, he analyses the functions of the building, studies the best way of building it, making use of both new and old materials. He accepts the new forms and effects that emerge . . .[5]

Here human action makes a surreptitious entrance through the back door. Although the architect only *accepts* the 'new forms and effects' rather than *designing* them, at least the forms and effects themselves 'emerge' from a human act of technological and functional analysis, where 'function' has been reduced in scope to mean something like 'expected activity patterns'.

Despite their complexity, however, accounts like this are still made up from combinations of simpler 'external forces' components. How credible are these, if we explore them in more depth? According to the simplest *Zeitgeist* version, people acting on and in the built world are subject to the laws of destiny, which work through them. As Oswald Spengler once put it,

> the Zeitgeist is inexorable; as inexorable now as it was in Roman times. There, too, it would have been absurd for a young intelligent citizen to hatch out some irrelevant variation of post-Platonic philosophy, if he was clever enough to organise armies and provinces, and to plan our cities and roads.[6]

Nikolaus Pevsner uses the same idea to explain why a dramatic new form of naturalistic medieval carving appeared with the building of the Chapter House at Southwell Minster:

> So we are left with the only explanation which historical experience justifies; the existence of a spirit of the age, operating in art as well as philosophy, in religion as well as politics. This spirit works change in style and outlook, and the man of genius is not he who tries to shake off its bonds, but he to whom it is given to express it in the most powerful form.[7]

Speaking of modernist rather than medieval design, Mies van der Rohe supports a similar view:

> I believe that architecture has little or nothing to do with . . . personal inclinations.

> True architecture is always . . . the expression of the inner structure of our time from which it springs.[8]

As these examples indicate, there is an ambiguity at the heart of most versions of the *Zeitgeist* problematic. On the one hand, to use Spengler's own terms, the *Zeitgeist* is 'inexorable'. On the other hand, however, it is merely 'absurd' (rather than impossible) to be a Roman philosopher. It is unclear, therefore, whether the *Zeitgeist* is supposed to dictate what actually happens, or what ought to happen, or elements of both.

This ambiguity leads to all sorts of dilemmas for those who call on the *Zeitgeist* problematic to explain their practical work in the form-production process. Here, for example, the American architects Peter Eisenman and Frank Gehry worry together about this issue:

> What I am asking is, do we have any right to say that what we believe to be the Zeitgeist suggests that it looks like our architecture as opposed to looking like eighteenth-century architecture?[9]

Because it cannot deal with worries like this, the credibility of the *Zeitgeist* approach has long been rejected in most intellectual circles. According to Heller, writing as long ago as 1952, the idea was already 'by common consent completely out of date'.[10] It is still, however, extremely influential both in popular culture and amongst urban change professionals (particularly architects), so we have to take it as seriously as we would any other practical problematic.

The reason why it is regarded as out of date by most thinkers nowadays, is that its validity cannot be tested against empirical evidence (as Eisenman and Gehry saw) because it forms a closed, circular argument. Why is Southwell Minster similar in style to other religious buildings of its time and region? Because of the *Zeitgeist*. But how do we know there is a *Zeitgeist*? Only because Southwell Minster is similar in style to other religious buildings of its time and region.

To point out that this forms a circular argument is not quite the same, however, as proving it wrong: nobody, surely, can ever prove or disprove the notion that destiny is at work in human affairs. Ultimately, the plausibility of such ideas must be a matter of personal belief or disbelief. They might be true, or they might not, but we can't tell because their 'closed system' character makes it impossible for us to check them against the evidence of what actually happens in the world.

Surely, however, few of us would find it plausible to believe that human actions have *no* impact on the form-production process. To the extent that we allow that they do affect it, the *Zeitgeist* approach still leaves us with no understanding of *how* they do. Relating only to those aspects of the form-production process which (were they to exist) would ultimately be beyond human intervention, the *Zeitgeist* approach can offer us no explanation which we might use in making better places.

Despite this crippling disadvantage from our point of view, however, the *Zeitgeist* problematic still raises an important question. Given that it was seen as 'completely out of date' by the early 1950s, how come it is discussed as a live issue by highly intelligent, well-known and fashionable architects such as Eisenman and Gehry some forty years later? The fact that, despite its intellectual shortcomings, it is still deeply embedded in design culture, strongly suggests that it must be playing some practical role at an ideological level. It is important to understand this role, if we are to understand why people think and feel about built form the way they do.

In the critique of the *Zeitgeist* approach sketched out above, we can see ideology supporting the current workings of the form-production process at three levels. The first involves claiming that the current form-generation process (or some selected aspect of it, as Gehry and Eisenman saw) *is* the *Zeitgeist*: that it has some special significance beyond being merely 'what happens to be done nowadays'. The second level involves the argument that this *Zeitgeist ought* to be followed: in other words, that current ways of doing things should be supported and maintained. The third involves the claim that the *Zeitgeist* is 'inexorable', so that any attempt to change the current character of form-production is doomed to failure from the start. This has a double ideological impact: at the same time as it discourages anyone from trying to change the current situation, it also effectively absolves practitioners (in their own eyes as well as others') from any blame if the results seem horrible to their users: 'it's not my fault, it is just in the nature of things that urban places are like that nowadays.'

In the end, then, our critique of the *Zeitgeist* problematic is not barren after all. It has revealed useful insights about the ideological dimension of the form-production process. At this point, though, we seem to have gleaned as much as we can from this approach. To get beyond its now-obvious limitations, we have to move towards other 'weaker' versions of the 'external forces' problematic which have some more manifest connection with human action.

There are various versions which appear to fit this bill, in the cultures both of architecture and of town planning – versions which see urban form as generated directly by social systems. This general approach has a number of different strands, each based on the view that some system, such as 'society' or 'the economy', 'needs' certain sorts of built forms in order to work efficiently.

The architect Richard Rogers, for example, gives a very clear example of an economistic version of this approach when he writes that, 'Most contemporary architecture is . . . the product of stark economic forces rather than the work of a designer, it represents the logical product of a society which sees the environment in terms of profit.'[11]

The problem here is that accounts like this regard particular people as mere pawns of social systems – 'cultural dopes', as the British sociologist Anthony Giddens puts it – rather than as 'actors who are highly knowledgeable . . . about the institutions they produce and reproduce through their actions'.[12] Given the wide range of forms which are constantly being produced within any given economic system, for example, Giddens's theoretical objection seems to be amply borne out in practice. Here again, it seems that the practical value of this 'structuralist' problematic must lie in its ideological effects rather than in offering any other useful insights.

Viewed in this light, it is not hard to see how this problematic, which conceives of social structures as disembodied forces external to the actions of particular people, supports the maintenance of the status quo in form-production terms. It deflects the potential for public discontent by obscuring the responsibility of real human actors in the production of unlovable urban places. It is not particular developers, or politicians, or architects or others who are to blame. No. It is 'society' or 'the economy', or 'market forces'.

The 'market forces' version of this problematic is particularly insidious, from our point of view, in its ideological effects. According to this view, when buildings are produced as commodities for sale, in what amounts to a retail market for urban form, the operation of market forces will ensure that only the ones which appeal to potential purchasers will make sufficient profits to be replicated by developers, thus ensuring that the results are 'what people want'.

There are, however, many flaws in this argument. First, it is perfectly obvious that most users are not involved in the purchase of all the buildings they use in the course of their everyday lives. Even people who own their own homes, for example, have to adapt their lives to space bought by others, in all the sorts of buildings they use for work and leisure pursuits. Second, the argument that consumers get what they want through market signals cannot in any case apply to public space, because public space is extremely hard to purchase as a commodity for sale. The point here is that cities exist for processes of communication and exchange between people – that is the only reason for having them in the first place – and public space is a key medium through which these processes take place. In capitalist contexts, it is easy enough to commodify communications media in general; as with the mail, the telephone, the fax and so forth. Still, face-to-face contact is crucial in many situations, and to achieve it there has to be an irreducible (and large) proportion of space which must be left as *public* if the city is to function effectively as a setting for communication and exchange. This space has to be 'freely' available, in all senses, to its users. In regard to this public space there can therefore be no capitalist market through which users can signal what they love and what they hate. The consequence of course, is well known: the concentration of design effort on the (saleable) building, and the reduction of the public realm to (literally) 'worthless' SLOBB: mere Space Left Over Between Buildings. The 'market forces' approach is another problematic whose prominence can more easily be explained by its ideological effects in supporting the status quo, rather than by its ability to explain how form-production works.

Let us now move on to the more complex 'natural necessity' approaches;

which stress the form-generating role of natural factors such as the proper-ties of climate and of building materials. Of the English cottage, for example, we are told that 'its designer is the English climate',[13] whilst more generally we hear that, 'for thousands of years wood and stone have determined the character of buildings.'[14]

Unlike the *Zeitgeist* approach, this 'natural necessity' problematic does not form a circular argument. We can test the validity of the 'problematic of climate', for example, against the evidence of what actually happens in the world, by examining whether places with similar climates do necessarily, in practice, have similar patterns of urban form. There is a great deal of evidence to show that this idea is unconvincing. For example, Amos Rapoport, discussing the Ona people of Tierra del Fuego, points out that, 'Although the climate there is almost arctic, and the ability to build well is shown by the presence of elaborate conical huts for ritual purposes, only wind-breaks are used as dwellings.'[15]

We might expect that the idea that climate determines form would be even harder to sustain in wealthy industrialised situations, because of the far wider resources available to designers. This seems to be borne out by the facts in many cases. For example, Le Corbusier's Mill-owners' building at Ahmedabad, India (Figure 1.1) – supposedly designed with great attention to climatic factors – looks remarkably like the same architect's Carpenter Visual Arts Center, at Harvard in New England (Figure 1.2), where the climate is utterly different. And in any case, as Reyner Banham pointed out long ago, 'we now dispose of sufficient technology to make any old stan-dard, norm or type habitable anywhere in the world. The glass skyscraper can be made habitable in the tropics, the ranch-style split level can be made habitable anywhere in the US.'[16] Whether we applaud or regret it (and I regret it very much, because of its disastrous ecological implications), we should have to be blind not to notice the relative uncoupling of form from climate which has taken place in much recent architecture.

In any case, there is a great deal of evidence that 'acceptable' standards of climatic comfort are themselves socially determined. This is often graphi-cally demonstrated, for example, in colonial situations, where architectural forms originating in the colonial power's home territory are sometimes built to symbolise the power of the colonial regime, almost regardless of climatic conditions. For example, the Cathedral of Notre Dame in Saigon, designed in 1875 by the French architect Jules Brouard, was built in a mixture of Romanesque and Gothic forms associated with the architectural history of France. This was consistent with the then-current 'assimilationist' colonial policy of the French in Indo-China, which for political reasons stressed the superiority of French culture over that of the indigenous people. In con-sequence, as the architectural historian Gwendolyn Wright tells us,

> no-one acknowledged that the hot, humid climate of Saigon required
> better ventilation than this structure, more suited to northern France,

Figure 1.1 Le Corbusier: the Mill-owners' Building, Chandigarh, 1954.

could offer. Two spires were added in 1894, but not until 1942 would
the French finally concede the need to pierce additional openings in the
lateral chapels, following more or less the traditional Vietnamese
practice for ventilating pagodas.[17]

Significantly, the new openings were pierced only after the colonial policy
had been changed, in an attempt to reduce the local hostility towards

Figure 1.2 Le Corbusier: Carpenter Visual Arts Center, Harvard University, 1965.

French domination, from assimilationism to 'associationism': a new cultural–political approach whose impact in Morocco was described in 1932, by the French architect Joseph Marrast: 'thus, little by little, we conquer the hearts of the natives, and win their affection, as is our duty as colonizers'.[18] Clearly, 'acceptable' standards of climatic comfort are far from absolute, even with respect to a given culture: in some situations, at least, they are politically determined. And in any case, there is considerable evidence that human actions have now begun to influence climate; this is not a one-way traffic any more, if ever it was.

It is clear from all these examples that, in both the vernacular and the modern situations, neither climate nor the need for shelter can be seen as direct determinants of physical form. They obviously place constraints on design, but the nature and strength of these are highly problematic. What, in a particular social situation, determines the standard of climatic comfort which people (and *which* people?) find acceptable? What determines the level of expenditure and the type of technology which is made available to achieve it? These are the sorts of questions that must be addressed by any credible explanation for the relationship between climate and physical form.

A second widespread version of the 'natural necessity' approach sees built form as generated by the physical characteristics of building materials. This problematic finds expression, for example, in Le Corbusier's view that reinforced concrete 'has brought about a revolution in the aesthetics of construction. By suppressing the roof and replacing it by terraces, reinforced concrete is leading us to a new aesthetic of the plan, hitherto unknown.'[19]

This idea holds very little water. Consider for a moment the vast range of different building forms which have been built in any particular material. What similarities are there, for example, between the massive stone Nativity Church at Bethlehem and the lacy Gothic of Milan Cathedral (Figures 1.3, 1.4). Or between the ways in which concrete is used by Robert Maillart (Figure 1.5) and by Le Corbusier himself in the examples we have already seen. The 'natures' of all these materials (if this is a useful concept at all) seem so adaptable that they can have little direct impact on form.

This is perfectly obvious to many thoughtful designers. The American architect Timothy Culvahouse, for example, reminds us that,

> Despite the different limits set by their manufacture, a sheet of stone veneer has more in common with an opaque glass spandrel panel than with a block of the same stone used in bearing, and is more like the former in size and shape. Similarly, rather than the differences between stone and brick requiring a difference of treatment, it is a common property of the two that they allow either diverse or similar treatment, as the occasion demands.[20]

Figure 1.3 The Church of the Nativity, Bethlehem.

Figure 1.4 Milan Cathedral.

An extension of the 'form is determined by materials' argument, probably with a wider following nowadays, ascribes a form-determining role to structural systems rather than to the materials used in them. Even when, like Timothy Culvahouse, they can see perfectly well that materials themselves have a limited role in the form-generation process, many architects still cling to some version of this structural determinism. As Culvahouse himself expresses it:

> Most building materials in fact have a wide enough range of use that they can as readily be shaped like one another as they can be shaped differently; the nature of the material does not demand differentiation. *The building system – frame and bearing being the two most basic types – typically determines the shape and joining of units.*[21]

This argument seems to have a certain plausibility. Surely, for example, the design of Robert Maillart's Cement Hall at the 1939 Swiss National

Figure 1.5 Maillart: Cement Hall, Swiss National Exhibition, Zurich, 1939.

Exhibition (Figure 1.5), designed to exhibit the structural potential of reinforced concrete, must truly be generated by structural necessity. Or is it? Giedion tells us that Maillart

> often designed his bridges with a single curve on a scrap of paper while en route from his Zurich office to his Bern office. The specialist's simple calculations would have been too insufficient a guide towards new solutions where invention in the fullest sense of the word, plays a more decisive part than calculation. It is significant of Maillart that he made his calculations a servant and not a master.[22]

Even in the case of Maillart – trained in an entirely 'hard' engineering background – it is clear that the designer first decided on a form, and only later checked its structural feasibility through those calculations which were 'the servant and not the master' of design. What we find is that the supposed generator of form really only puts constraints on the possibilities which are open to the particular designer. We also find that technological standards are, in the end, socially determined: the structural engineer Anthony Hunt, for example, has said that the 'advanced' structural system of the Inmos factory cost some 30 per cent more than the simplest system which would have performed the same practical task, because of the client company's desire to use the structure of the building to symbolise Inmos's high-technology role in the production of microchips.[23]

It seems clear from these examples that structural determinism, as an explanation of the form-generation process, ultimately falls apart. It is certainly true that consideration of what is structurally possible (which is often – perhaps even usually – an economic matter) places strong constraints on the form-generation process. In some cases, a designer's understanding of how structures work may even suggest a particular form. But all this is a far cry from structure as a generator of form in its own right.

In any case, we are moving here into the realm of technology – the human manipulation of natural potentials – rather than considering the impact of natural properties in themselves. The idea that technology determines built form constitutes a third problematic in its own right, often used to make sense of what happens at all physical scales.

At the largest scale, this approach is often used to explain the forms of the communication spaces which bind urban form together; according to this problematic, these spaces are determined by technologies of movement. Nobody could sensibly deny that the availability of the car, for example, has been associated in some way with changes in urban form: 'The auto-mobile split the city open at the seams, and to this day we are frantically trying to hold it together with patching on a worn-out fabric', as the architect Arthur Gallion and the planner Simon Eisner despairingly say.[24] The point is, though, that all sorts of patching operations have been carried out in different places at various times, and these have interacted with different levels of car-availability to generate different patterns of urban form at various scales. The availability of the car makes possible certain kinds of low-density, dispersed forms which were impossible before it. But that is far from saying that these changes were caused by the car. People with the power to do so decide how (or whether) to use the potential which the car offers; and the choices they make are shaped by all sorts of factors other than the car's own availability.

Notable amongst these factors is the level of state intervention in the urban transformation process. Strict 'green belt' policies in many British cities, for example, have contributed to the production of urban patterns which are markedly less dispersed than those in the USA. Differences in state policies have their effects at the smaller scale too. British government policies on road safety, for example, have changed over the years, and urban form has changed with them; from the open grid layouts of the 1930s, to the hierarchical culs-de-sac of the 1970s and after. Clearly, at all scales, it is the actions of powerful people and the policies of powerful institutions, rather than the availability of the car (or any other form of communication technology) in itself, which have altered urban form.

At a smaller physical scale, technology is often held to have generated built form through the increasing mechanisation of the building process: 'mechanisation takes command', in Siegfried Giedion's famous phrase.[25] The idea of mechanisation as a disembodied process with its own 'needs' is no more convincing at this smaller scale, however, than it was at the larger scale of transport structures.

It is certainly true, at least in the industrialised world, that an increasing proportion of buildings has incorporated an increasing proportion of machine-made components over the last century or so. In many cases, particularly in recent times, the buildings made from these components have been simple, even 'minimalist' in form; but that is not entailed by the process of mechanisation itself. Many of the minimalist icons of modernism, often thought to express the industrialisation of the building process, were in fact constructed using mainly traditional techniques: I well remember the shock with which I discovered that Gerrit Rietveld's Schröder House was in fact built up from rendered brick (Figure 1.6).

Conversely, one of the earliest applications of machine-made components in buildings, for example, was far from minimalist in its effects: it involved the substitution of machine carving, moulding or pressing for handwork in the late nineteenth century. To reinforce the point, this approach has been revived more recently with, for example, the production of complexly modelled components in materials such as glass-fibre and glass-reinforced

Figure 1.6 Gerrit Rietveld: Schröder House, Utrecht

concrete: mechanised building processes do not in themselves imply any particular vocabulary of form. As we saw at the larger scale, technological developments have created opportunities for new kinds of forms, but there is nothing inevitable about their practical outcomes. The opportunities have been seized on by particular people, in particular relationships of power, for specific purposes in specific situations.

The problematic of technology, then, does not explain anything in itself; but it certainly draws our attention to the importance of technological developments in opening up new possibilities in the production of urban form. Perhaps more importantly, it also raises a number of key issues of action, culture and power which any adequate explanation of the form-production process will have to address. Let us explore the extent to which these issues are illuminated by our final strand of the 'necessity' problematic: the idea that form is determined by the uses or 'functions' which it has to house. This notion, encapsulated in famous phrases like 'a house is a machine for living in',[26] is somewhat out of fashion with advanced architectural thinkers at present, but is still implicit in the attitudes of many who work in the form-production process.

At the larger scale of urban form, this approach to explanation is not very plausible. We can start to see why if we consider the relatively simple example of a small 'vernacular' settlement in Egypt. The development of the village began with a couple of small dwellings located near to a well. The 'functional' explanation – the location of the well determines the location of the dwellings – has a certain plausibility in relation to the earlier stages of such a settlement; though even here there are clearly alternative positions where the first dwellings could have been located, all still 'functionally' close to the vital water source. As soon as the first houses are positioned, however, other factors come into play. Later buildings inevitably impinge on those which are there already and, as the Egyptian architect Amr El Sherif convincingly shows,[27] the positioning of these later arrivals, to form the larger village layout, takes place according to broader social factors which are not merely 'functional' in nature. These factors include each individual's duty not to damage the neighbour's rights of light and access, for example, which are codified in Islamic law.

What we see here is a complex set of social, cultural and religious factors, which can only be compressed into the rubric of 'function' if that word is allowed to expand its meaning to a level of generality which robs it of all explanatory power. And even then, the complex factors codified in Islamic law do not determine form by themselves; for the same code of law is associated with the production of quite different patterns of built form across the breadth of the Islamic world.[28] We surely do not need to multiply examples here to see that 'functional' explanations of urban form are even less plausible in complex modern industrial societies.

The functionalist problematic is no more convincing at smaller scale, for there seems to be no evidence that architectural form is directly generated by the activities in and around buildings or outdoor spaces. For example, take the favourite café in which ('unfunctionally', I suppose) I like to write. Here is what it says on the back of the menu:

> The Congregation House was built in the early 14th century and was used by the University for its congregation and council, and at other times by the Rector and parishioners of St Mary's Church. In 1480, it sank into neglect and became a store house for university lumber.
> Later still it was used to store the University fire engine and continued to be a 'fire station' until 1871. Whereupon it was used as a chapel for the new non-collegiate students.
> In the years following it was used as a Brass Rubbing Centre before being converted, finally, into the Convocation Coffee House in 1990.[29]

Only the age of the building makes this history of changing 'functions' in any way unusual. When cement silos have been reused to make a high style architect's office, gun emplacements have been designed in the form of thatched cottages, a Malaysian shopping centre has been taken over as a university and a Russian nuclear power station has become a vodka-bottling plant,[30] it is surely obvious that built form cannot plausibly be seen as 'generated' by the activities it houses – a point which becomes more and more obvious to perceptive designers like Bernard Tschumi:

> Whether cultural or commercial, programs have long ceased to be determinate, since they change all the time – while the building is designed, during its construction and, of course, after completion. (At the Parc de la Villette, one building was first designed as a gardening center, then reorganized as a restaurant by the time the concrete frame was completed, and finally used – successfully – as a children's painting and sculpture workshop.)[31]

This point gains further support from the fact that buildings which have been converted from one use to another are sometimes preferred, by their users, to others which were purpose-built. 'Loft living', where in certain circles it is the height of chic to live in a converted down-town factory or warehouse, is a celebrated case in point. No doubt the 'down-town' location is relevant here, but the loft's particular visual character and historical associations are also important, as its frequent use in films and in advertising sets so amply testifies.

The potential for issues of image and historical association to override considerations of practical 'function' is not confined only to the members of a cultural elite. In a revealing example, the Italian sociologist Maria Vittoria Giuliani, studying the conversion of a complex of sixteenth-century religious buildings into social housing in Rome, found that residents were generally very satisfied with the dwellings which had been made, despite the fact that their rooms were so unusually shaped and appointed that one bedroom contained a (non-functioning) fountain. Questionnaire evidence suggests that practical inconveniences were overriden by the fact that residents found that the designs provided wonderful talking-points when visitors came.[32]

The overriding of brute 'function' by issues of historical association can go further, in certain circumstances, to become a matter of political policy. In the period after the French Revolution, for example, it was French government policy to put the religious and aristocratic buildings of the *ancien régime* to new revolutionary uses, to symbolise the triumph of the new order. For a while at least, France underwent a kind of architectural cross-dressing. To explain in 'functionalist' terms why law courts took over buildings designed for religious worship, we would once again have to allow the word 'function' to expand its meaning to encompass everything and therefore, ultimately, nothing.

All this is not to say, of course, that there is no relationship between physical forms and the patterns of activity which they house. The point is merely that this relationship is a loose one, in which some activities might be impossible within a given form, but many others can probably be carried out perfectly well. Indeed, it is increasingly often argued that built form should be designed from the outset to facilitate a variety of different 'functions'.[33] In the design guide for the renewal of the Hulme area in Manchester, for example, architects are instructed to design all buildings in such a way that they can easily be taken over for future housing use.[34]

Looking at this the other way round, it is equally clear that in practice it is almost impossible to design any form to support only one functional pattern. Nineteenth-century Panopticon prisons were 'scientifically' designed as physical machines for subjecting their inhabitants to the most rigidly disciplined and surveilled pattern of activity.[35] But Abingdon Gaol, designed on this pattern, now functions rather well as a centre for various kinds of leisure activities. Even a prison needs its gaolers to imprison those within.

Even at the smallest scale of individual rooms, the idea that particular functions need particular spaces and forms closely tailored to them was long ago discredited by Cowan. His study of how rooms are used in hospitals implies that the large majority of activities can all take place within a room of normal domestic ceiling height, about 14 square metres in area.[36] Musgrove and Doidge, in their study of room classification at University College London, show that the greatest practical inhibition against using a room for many purposes is often the label on the door, which tells potential users that the space is meant for some particular purpose.[37]

Even with commercial offices, where we might expect the most rigorous attempts at economics-driven 'functional' efficiency, the architect Frank Duffy found that the symbolic capacity of office layouts to express values such as status, is greater than their capacity to achieve operational results such as more or less internal intercommunication.[38] At a still smaller scale, Reyner Banham pulls the flush on the functionalist idea when he points out the wide range of functions which even simple artifacts like chairs can perform:

> Not only are they bought to be looked at as cult-objects, they are also used for propping doors open or (in French farce) shut. They are used by cats, dogs and small children for sleeping in; by adults as shoe rests for polishing or lace-tying. They are used as stands for Karrikots and baby baths; as saw horses; as work benches for domestic trades as diverse as pea-shelling and wool-winding; and as clothes hangers. If upholstered and sprung, they can be used for trampolining practice; if hard, as bongo drums. They are persistently employed as step ladders for fruit-picking, hedge clipping, changing lamp bulbs and dusting cornices.[39]

Not only does function not generate form, but we are also faced with the problem of understanding how our own ideas of 'function' are themselves generated.

In practice, designers' definitions of 'function' tend to home-in on whatever seems to be the most predictable and 'legitimate' range of activities for the place which is being designed. Many other activities which particular users might desire are, of their nature, never taken into account: which designer, for example, would consider the dimensions of a bedroom cupboard in terms of its suitability for concealing an illicit lover? And what of the many other, more 'legitimate' situations where one person's 'functional' is 'non-functional' for others? Think, for example, of Frank Lloyd Wright's decision to place the door handles in Tokyo's Imperial Hotel very high up the doors because he liked what, to him, was the graceful sight of diminutive Japanese chambermaids stretching up to reach them. Or, if you want a less recherché example, think about the functionality of 'traffic calming'. Here functional conflicts, and consequent perceptions of danger, achieve an environment which is better for pedestrians, in functional terms, because vehicles are 'unfunctionally' forced to move more slowly. '*Whose* function?' is a question which inescapably presents itself at every turn, in the real world, unless it is obscured by some sort of ideological veil.

Any useful explanation of the production of built form will therefore have to treat as problematic both the concept of function itself, and its relationship to built form. Which factors decide how 'function' is defined in particular instances? And whose interests are served by designing the built environment to suit one pattern of uses rather than another? However these questions are to be answered in particular instances, it seems clear from all these examples that 'function' is itself a humanly-constructed concept. In the end, human purposes somehow creep back into all these functionalist accounts of the form-generation process. People *make* the urban environment, after all. It never just *happens*.

To summarise the arguments of this chapter overall, it should by now be clear to us that all the supposedly 'natural' forces which impinge on the form-production process are mediated through human intervention. Even processes of physical decay, brought about by the direct action of wind, water or plant life, require the collusion of the humans who neglect to carry out the maintenance required to prevent it. It should also be clear, however, that this does not deny the importance of natural factors such as climate, or the physical character of building materials, in the form-production process. It is merely to point out that they are important, not as generators of built form in their own right, but rather because they offer potentials and impose constraints on the forms which human action can generate in particular situations.

The examples in this chapter have also suggested that human action in the form-production process depends both on the objectives of the actors concerned and on the resources available to them in pursuing these objectives. And, as we have seen, both the objectives and the levels of resource-availability are embedded, at different levels, in wider political and economic systems. Of course, in particular situations, particular actors might want to 'go with the flow' of natural factors – to minimise the alteration of natural land-form, for example, or to seek forms from nature as 'inspirations' for design. But then again, they might not.

All this means that whenever we come across claims that 'natural forces' generate designs, alarm bells should start ringing in our minds. Claims like these are always either untestable or untrue, and we have to ask ourselves whose interests might be served by advancing them: as we have seen, there is an ideological dimension to all this. A claim that *this* (whatever it is) constitutes the Spirit of the Age, for example, acts as a rallying cry for others to follow the same path which *this* takes. Or a claim that *that* constitutes 'function' persuades others (and, perhaps as importantly, oneself) that *that* should be given a high priority in the form-production process. And in both these claims, the belief in 'natural forces' also absolves producers from any responsibility they might otherwise bear for the production of unloved urban places: 'please, God, it wasn't me, Function and the *Zeitgeist* did it.'

It is now surely clear that any useful explanation of the form-production process would have to put human action at its centre. That kind of problematic is what the next chapter is about.

2   Heroes and servants, markets and battlefields

The last chapter pointed towards the importance of human action in the making of urban form. To understand the form-production process, we need an approach which takes account of real people doing practical things. In this chapter, we shall review a range of problematics which might offer what we need here, since they all stress the role of individual action in the form-generation process.

The simplest of these 'problematics of action' is that of the 'heroic form-giver', which is based on the idea that built form is generated through the creative efforts of particular individuals. These heroes propose forms, whilst others merely implement them. More complex is the view that there are many actors involved in the form-production process, and that the outcome is determined by power-plays between them. The most basic version of this approach claims that those actors with the most power simply issue orders to those with less. More complex is the 'market signals' perspective – a more action-orientated conception than the basic market problematic which we already explored – which sees resource-poor actors such as designers responding to market signals which indicate the kinds of schemes which those with the necessary resources are willing to fund. An alternative, more sophisticated version is the 'battlefield' problematic, in which the various actors are seen not merely as ordering each other around, or as responding to market signals, but rather as plotting and scheming to use their power in the best ways they can devise, in attempts to achieve the built forms they want. In this chapter, we shall review each of these problematics in turn. As our starting point we shall take as 'hero' the architect: in the current complex division of labour, architects are highly visible at the sharp end of the form-production process as a whole.

Both in popular and in professional culture, it is certainly the architect who is most often cast in the leading role. In popular culture, this position is most famously celebrated in Ayn Rand's best-selling novel *The Fountainhead*. Still in print in many languages, the book has sold some nine million copies since its publication in 1943;[1] whilst the film of the book, starring Cary Grant, is still shown from time to time on prime-time TV. *The Fountainhead* is not entirely ignored in professional culture either: its fiftieth anniversary was marked by discussion in *Architecture*, the journal of the American Institute of Architects,[2] whilst Dana Cuff, a highly perceptive sociologist of the architect's world, tells us that her own first innocent views on architecture came chiefly, perhaps, 'from Howard Roark, hero of Ayn Rand's novel *The Fountainhead*, who pursues at all costs his personal vision in the face of society's mediocrity'.[3]

Throughout, the novel celebrates the idea that the prime generator of built form is the creative power of the individual architect. Of course, it is admitted, many other people are in various ways involved in the making of a building; but it is only the individual architect who breathes form into the process. In the words of the architect hero:

> Every creative job is achieved under the guidance of a single individual thought. An architect requires a great many men to erect his building. But he does not ask them to vote on his design . . . An architect uses steel, glass, concrete, produced by others. But the materials remain just so much steel, glass and concrete until he touches them.[4]

This view, in more measured guise, is deeply embedded in professional design culture too. It is expressed, for example, every time an architect refers to 'my building'. And it is reinforced, and disseminated to a wider public, through all those coffee-table books with titles like *The Buildings of Joe Bloggs*.

This idea that the individual architect has a crucially important influence on built form seems to be supported by a great deal of evidence. It is clear, for example, that certain architects do have remarkably distinctive and consistent personal styles, which mark their designs out from those of other people. Buildings by Le Corbusier, say, at a given stage in the development of his complex career, do have an obvious family resemblance between them; and they do all look different from those by (say) Daniel Libeskind or Aldo Rossi.

In practice, however, relatively few architects seem to play Roark's 'form-giver' role, though many more would probably like to: indeed the rarity of genius is a key message of *The Fountainhead* itself. In some way, a relatively small band of designers seems to mark out a fairly limited range of design paths, which are then followed quite closely by the majority of practitioners. There are so many 'followers' that the similarities between different individuals' designs, at a given historical moment, usually seem far more striking than their differences. It is only this, indeed, which enables us to talk in terms of 'architectural styles' as we do.

One explanation for these similarities puts them down to psychological differences between individual designers, seeing variations in creativity as the factors which distinguish 'leaders' from 'followers'. This may help to explain why most designers follow paths laid down by others, but by itself it cannot explain why those particular leaders are chosen as the ones to be followed along those particular paths at particular historical moments. If creativity alone were the key to architectural leadership, then (for example) the Archigram designers of the 1960s, like Ron Herron who made way-out proposals for cities which walked about, would have been assured of a massive following. In fact, their practical influence on built form was virtually nil. Creativity, it seems, is not enough. There must be other factors at work in deciding who follows whom.

The Fountainhead itself gives us clues about what these factors might be. Though poor Roark works desperately to maintain his heroic vision of his ideal creation, in the end he is driven to blow the building up because it is so botched by others' actions. This is all a matter of power. Roark has the power to destroy, but lacks the resources which he needs to turn his vision into bricks and mortar, whilst those who do have the resources which are necessary to build also have their own agendas about how these resources should be used. In some circumstances (probably rather few, as Roark found out) this agenda might be centred on a desire to support the architect in creating a work of art; but equally it might not. Anyone with any experience of the real-world development process knows very well that usually it is not. In most cases, therefore, the idea that built form flows directly from the architect's individual inspiration has to be understood as a powerful myth, rather than as a statement of fact.

Given the complex division of labour in the modern development process,[5] together with the fact that power is very unequally distributed amongst the various actors involved, it is equally implausible to think that any other actor, alternative to the architect, might be a heroic form-maker either. This raises an interesting question: how could such an implausible concept as the 'heroic form-maker' ever have become so widely accepted? What do any of the actors in the form-production process have to gain from it?

For architects, trying to make a living, the benefits are obvious. If form is believed to be the result of their own creativity – 'my building' – then it is theirs to sell in the market. As Ayn Rand said of Roark 'the materials remain just so much steel, glass and concrete until he touches them. What he does with them is his individual product and his individual property.'[6] At another level, this ideology also supports the interests of other powerful actors in the development process, for it implies that it is only architects who can be blamed for the creation of unloved places – these are *their* creations, after all. This is extremely convenient for everyone else involved, for it draws a veil over their activities, inhibits any deeper criticisms of the form-production process, and thereby enables it to continue unchanged.

In the end, then, the 'hero' problematic has helped us to develop our understanding of how the ideological level of the form-production process works, but it clearly has so many drawbacks in other areas that it can be of no further help to us. To make further progress we have to go beyond it, to explore the range of problematics which comprehend built form as generated through a process of interaction between a range of actors, each with access to different sources and levels of power.

The simplest way of conceiving these interactions is in terms of 'masters and servants', whereby those with the most power simply command the actions of those with the least. This concept is widespread, in both popular and professional cultures. In its commonest formulation, it is those with economic power – those who fund building projects – who are seen as ruling the form-production process, in a built-form version of 'whoever pays the piper calls the tune.'

At first sight, this approach seems very plausible. Buildings are extremely expensive to produce, and it seems likely that those who are able to put resources into developing them would do so for their own purposes. In the context of capitalism, these purposes are usually concerned with making profits: there is no reason, after all, to think that most major property developers are particularly interested in art for its own sake. The example of Lord Peter Palumbo, the commercial property developer who has also served as Chairman of the Arts Council, is surely the exception whose very noteworthiness proves the contrary general rule. And even if this were not the case, most major property developers themselves have shareholders, who invest their money in the expectation of profit, and will stop supporting the developer's very existence as a developer if acceptable profits are not produced.

To an important degree, then, the resources needed to construct buildings are only made available with strings attached. In their hearts, most people – including artist-architects – know this perfectly well. It was these strings which tangled up the efforts even of the heroic Howard Roark in Ayn Rand's novel, causing him such distress that he blew up his botched creation in a rage. But real-life architects have mortgages to pay and children to feed like anyone else. Like most of us, they would probably come round in the end to doing what their patrons dictate, without Howard's heroics.

On the face of it, all this supports the idea of the developer as master of all the actors in the development process; which offers one way of explaining, for example, why so many different architects are involved in producing similar buildings. If all office developers (say) have the same profit-orientated objectives, and if it is mostly these objectives which determine the forms of office blocks, then it is not surprising that so many office buildings are so much alike.

In practice, however, things cannot be so simple, if only because there are other contenders for a 'master' role. For example, the idea of 'the planner as master' seems to be widely held amongst architects, or architecturally-trained critics like Martin Pawley:

> Planning officers too have taken upon themselves more and more of the effective decision-making about the content and appearance of building projects. In architectural practice the consequences of disputes with planning and building control officers, in terms of delay alone, can be serious enough to force almost any compromise.[7]

This idea that 'the planners' can more or less dictate the outcome of the form-generation process does not usually seem to be borne out by the facts. In negotiation over the form of the Lloyds building in the City of London, for example, the local planning authority's officers initially pushed for an image reminiscent of the nineteenth century, but the final result had nothing of this character (Figure 2.1). In contrast, the layout of Oxford's Gloucester Green development (Figure 2.2) is remarkably like what was called for in the planning authority's urban design brief (Figure 2.3). Some you win, it seems, and some you lose.

In the end, then, this 'master and servant' problematic is no more convincing than the idea of the heroic form-giver. This is largely because it ignores the problems which face even the most powerful actors when they try to control the work of all the specialised experts, such as architects, who are involved in the modern development process. The idea that 'whoever pays the piper calls the tune' is unconvincing in this situation, because building projects are not at all like tunes. Tunes, after all, are predictable and are known in advance to those who pay the piper, whilst buildings are often one-offs, and in any case are always on unique sites. Even when buildings themselves are standardised (as, for example, in many speculative housing

Figure 2.1 Richard Rogers: Lloyds Insurance Building, London, 1986.

Figure 2.2 Mixed use development at Gloucester Green, Oxford: Donald Kendrick
        Associates, 1990.

estates) the overall layout of each 'estate' is unique. Patrons, therefore, cannot
know exactly what they will be paying for, in advance of some design process
carried out by design experts.

   As with the 'heroic form-maker' approach, the widespread nature of the
idea that 'those with the most power always win' probably owes a great deal
to its ideological role in the form-production process; it absolves those actors
who lack access to economic or political power from struggling too hard to
achieve whatever they believe to be the best form for the situation in hand.
Though sometimes a mite depressing, this makes the working lives of
relatively powerless actors a great deal less stressful and, no doubt, more
efficient for developers in economic terms.

   If patrons cannot know, in advance of the design process, exactly what
they will be paying for, they can nevertheless know whom they are paying
for, when they buy the services of professional advisers in the market place.
Clearly, patrons are most likely to buy the services of those whose track
records demonstrate a willingness and aptitude for working in the patron's
interest. As Phillippo advises, in a guide for developers: 'It is not advisable
to try to change the style of an architect; but to find an architect who in the

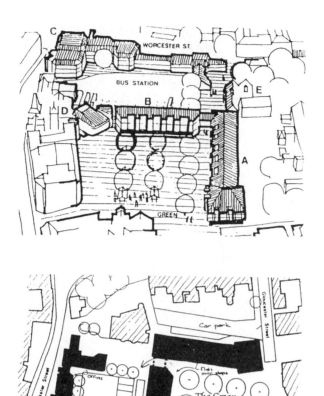

Figure 2.3 Oxford City Council: Gloucester Green Design Brief.

opinion of the market analysts is in demand.'[8] Advice of this kind is well-heeded in practice:

> Builder Erwin S. Wolfson of Diesel Construction Company concedes quite frankly that the reason most New York office buildings are designed by three or four firms is that these architects know exactly the sort of cost-saving structure that the owners want. 'And it's not the architect's fault that we don't get more distinguished buildings', Wolfson says, 'It's ours.'[9]

At this point, then, it begins to look as though the limitations of the 'master and servant' approach might be overcome if we adopt the perspective of the 'market signals' problematic, according to which the various actors in the development process are kept in line not by brute force, but by market signals which indicate the sorts of services and forms which patrons are willing to fund. How far can this can take us?

We can best explore this approach by starting from the position of design professionals setting out their stalls in the marketplace, competing with each other to sell their services. One important way of improving cost effectiveness in professional work is through a process of increasing specialisation, in which broad and complex tasks are split down into ever-narrower parts. For a given cost, this enables a greater degree of specialised expertise to be applied to each given aspect of the development process, to the benefit of whoever is in a position to buy the services concerned. The market process therefore supports the emergence of an ever more complex and specialised division of labour in the production of urban space.

Particular professionals, with their particular specialised skills, build up track records, for which they are hired (or not). Market signals, broadcast through the professional media, enable their competitors to see which sorts of services and forms are bankable, and which are not. In capitalist situations, disciplines of unemployment and bankruptcy ensure that it is only the bankable services and forms, and the ideologies which support them, which are widely replicated. Through their own responses to market signals, therefore, most of the actors in this complex division of labour get themselves into line, with no need for overt shows of force. All this makes sense in the abstract, but it ignores the practical difficulty of controlling all the experts in the development 'team' once they have been hired. Believe me: as an ex-property developer myself, I know how difficult this is.

The difficulty arises, at least in part, because of a mutual ignorance and antipathy between the various members of the development team, a state of affairs which arises through the process of increasing specialisation itself. As each new service is offered as an innovation in the marketplace, it has to be seen by potential buyers as being distinct from the other services which are already on offer: it has to develop its own 'unique selling point'. This means that the promoters of each new service have to emphasise the differences between that service and its possible competitors.

One result of this process is that many of the actors in the development process carry out their work according to different value-systems. The British economist Ralph Morton, for example, points out that, although architects and structural engineers both emphasise the fundamental importance of design in their work, 'they each mean by "design" an almost totally different activity', with architects stressing the art dimension and engineers the scientific. In contrast to both, he argues, surveyors often have 'a primary concern with market efficiency and value for money where value itself is defined in monetary terms'.[10]

Within this general situation, there seems to be a particularly strong conflict between the values of most architects and those of many of their patrons. In the context of capitalism, where most buildings are produced in speculative markets, and many patrons' objectives are primarily financial, we find that many architects nevertheless have non-commercial values. Comparing his own results with those of Anastasi, Mackinnon, for example, showed that US architects were far less motivated by financial considerations than ordinary citizens, let alone (presumably) property developers.[11] Interestingly, the divergence is particularly marked in the case of the prestigious 'leaders' of architectural culture.

Mackinnon carried out his study in 1962, but the situation is probably not changing very fast. In a 1990 review, Ralph Morton shows how limited is the teaching of economic matters in UK schools of architecture, suggesting that this 'seems to stem partly from a belief that the subject is peripheral and there is simply no time for it; but it stems also from a fear that contact with the philistine world of the economist will contaminate the creative imaginative world of the young architect'.[12] This situation is further complicated by a range of studies which clearly show that there are also considerable divergences in the evaluation of urban places between architects and non-architects[13] and between architects and town planners.[14]

At first sight, this situation seems fraught with potential disaster for all concerned. Patrons do not have the knowledge to design buildings themselves, whilst their professional advisers are difficult to control and also – particularly in the case of the architect – are often actively hostile to the sorts of objectives which many patrons have, in so far as they understand them at all. Morton himself certainly sees this as a negative situation, remarking on 'the failure of the built environment professions to use their enormous collective skills and knowledge to a common purpose'.[15]

All this means that it is extremely difficult for patrons to control the experts' work in any detail. Even if the patron and the design professional were to share the same objectives, so that the professional was consciously trying to implement the patron's stated policies, still a degree of autonomous professional action would in principle be unavoidable, because no policy can ever be stated in a form which is detailed enough to be directly applicable, without interpretation, to every individual design situation. In the real-life situation of the development process, where the objectives of the patron and the expert are in conflict, it is even more difficult for the patron to exercise close control, as Dietrich Rueschemeyer reminds us:

> Where complex knowledge is used in the performance of work . . . it makes control and supervision very costly if not impossible since detailed control of experts requires equally well-qualified controllers. 'Lay' customers – however rich, prestigious or powerful – cannot themselves exercise control because they often do not know enough even to define what their problem is, not to mention monitoring its solution.[16]

This general point is certainly relevant to the form-generation process in particular, as the architect Vittorio Magnano Lampugnani points out:

> The public client is almost without exception an abstract entity, a vacuous and vague sort of bureaucratic figment. And the private client, who at least puts in an appearance – usually – in flesh and blood, is not capable of expressing precise, concrete and unequivocal demands.[17]

The sociologist Robert Gutman also supports the view that many patrons cannot specify what they want with any certainty:

> I have spoken with architects for several of the universities involved in major building projects here [Britain] and in America, and they are agreed that the task of developing university briefs was difficult but also fascinating and exciting. It was difficult because no one involved in the clients' organization – not the vice-chancellor or president, not the building committee, the department heads and professors – no one was able to articulate for them in any easy fashion their objectives except in the most vague terms.[18]

Admittedly this was written back in the 1970s, but more recent opinion confirms the same view. Bernard Tschumi – who, as an eminent practising architect, should be in a position to know from first-hand experience – tells us that 'in our contemporary society, programs are inherently unstable.'[19] As he sees it, 'Few can decide what a school or a library should be or how electronic it should be, and perhaps fewer can agree on what a park in the twenty first century should consist of.'[20] This inherent vagueness gives any expert actor a degree of autonomy, which can be enhanced by drawing on the power of inner resources such as initiative, determination or moral commitment, rather than merely relying on access to external sources of economic or political power. When we start focusing on such potentials, however, we have moved outside the limits of the market problematic. We now find ourselves in a place which resembles a battlefield rather than the friendly bustle of a marketplace. How far can this 'battlefield' problematic offer us further insights into how form-production works? In particular, how far can it help us understand the scope for using the relative autonomy of particular actors to outwit the big battalions?

A particularly adroit example of how to play a weak hand with consummate skill is given by the architect Zaha Hadid, designing a housing project for the IBA organisation in Berlin. Here, Hadid has been asked to design a three-storey building, but she does not want to do so:

> I always made faces and frowned, so they said mine could be five storeys. So I asked was that an average? I spoke no German, which is a good thing sometimes. I don't speak Japanese or German, so I can always

pretend that I don't understand what they are saying. They always say
we didn't understand what you asked us so the contract is wrong, and so
on. So I played the same game. I asked was it an average of five storeys?
And they said yes. After many trials and errors we had two buildings
. . . one is eight storeys high and one is three, averaging out five and a
half. So I had half a story [sic] to bargain for. Again that was crazy, but I
said you did tell me in writing it was an average of five and that was
that, as far as I was concerned.[21]

In this situation, Hadid has very little obvious power in the 'master and
servant' sense, yet she has achieved more or less what she wanted, through
an adroitly handled process of negotiation. If, as relatively powerless
people, we want to maximise the impact we can make on urban form,
there is much we can learn from this. Let us analyse the situation in more
detail, to see if we can get a clearer understanding of how she did it.

First, she has what Shoukry Roweis[22] calls 'knowledge power': she knows
things the others do not, and the others need that knowledge. She has
something they want, which gives her an initial bargaining position.
Second, the strength of this position is enhanced by the fact that Zaha
Hadid has a considerable international reputation in the world of avant-
garde architecture. This endows her with what the French social anthropologist Pierre Bourdieu calls 'cultural capital';[23] which, no doubt, is
amongst the reasons why she was hired in the first place. The logic of their
own commissioning decision implies that the people who hired her *must*
respect what she says and does. Third, the division of labour in the modern
form-production process is organised in such a way that it is usually only
'designers', such as architects, who make proposals for physical designs,
except in the most general terms. As an architect, this gives Zaha Hadid a
crucial element of initiative, so far as physical form is proposed. It is her
proposal, once made, which sets the agenda for the subsequent process of
negotiation about form. Taken together, these factors of knowledge power,
cultural capital and initiative give Hadid three cards to play in the negotiation game. Clearly she must have played them well, since she gets what she
wants. How does her strategy work?

First, we can see that these cards are not played indiscriminately, but are
mobilised in support of clear objectives. We can see, for example, that she
has a clear conception of the form – or at least, the kind of form – she wants.
This conception is set according to her own internal rules. Partly these are
rules about what constitutes 'good design' in her terms, but she also has
rules about the way architects ought to behave in their relationships with
their clients, rules about which potential negotiating ploys would be
legitimate in which circumstances. If these internal rules are transgressed,
internal psychological sanctions come into force. Beyond a certain point of
compromise over the physical form, for example, a sense of guilt – a sense of
betraying one's own values – might have become so strong that quitting the

job might have been the only way of coping with it. In our example, this point was never reached. We might sense a degree of guilt, however, about working to the letter rather than the evident spirit of the client's brief – a sense of guilt warded off by reference to another, 'higher' internal rule of general fair play. '*They* always say we didn't understand . . . so I played the same game.'[24]

Not all the rules and sanctions within which Hadid has to work are internal ones. There is also a complex envelope of external rules and sanctions which determine the space within which she can operate. For example, too much design compromise – even if she could live with it herself – would run the risk of losing the cultural capital which is bestowed by the acclamation of her peers. This would be a serious matter, for cultural capital brings with it respect, and therefore enhanced negotiating power. Conversely, pushing too hard to get what she wants runs up against the ultimate external sanction of unemployment. Repeated too often, this would lead in turn to the higher order sanction of bankruptcy. Together, these webs of internal and external rules, and the sanctions through which they are enforced, create a 'field of opportunity' within which the designer can work. The success of the negotiation, from her point of view, depends on her ability to get where she wants to within this field. And that, in turn, depends on her ability to mobilise her own resources – resources such as initiative, determination, knowledge and cultural capital – so as to influence the other parties to the negotiation in the most effective way.

The effective targeting of resources depends largely on mobilising them to offer the other actors things they want, or to prevent them from getting these, unless they grant one's own objectives. The practical difficulty here lies in knowing how far the other actors can be pushed before they arrive at the limits of the opportunity field, where they come up against internal or external sanctions on their own actions.

For example, developers working in the private sector have rules about making profits. These are not optional rules, for they are externally enforced through sanctions of bankruptcy; in a capitalist society, private-sector developers have no escape from this, if they want to stay in business. But where does the limit of the field of opportunity lie in this regard? How does the designer (for example) know how far developers can be pushed before they really have to dig in their heels? It is not hard to see that the more the designer (or any other actor) understands the rules and sanctions of the other actors – particularly those with the most power – the more effectively the designer's own resources can be targeted. In this particular case, for example, it would clearly be advantageous for designers to understand how to do developer-type financial feasibility calculations, to prevent the wool being pulled over their eyes too easily.

In the Zaha Hadid example, she is in fact negotiating with a non-profit developer of social housing, so different rules and sanctions apply. Still, even developers like these have rules about how their resources are to be allocated, and targets about how many housing units (for example) they are to build for a given allocation. Sensibly, she accepts these limits; arguing about the form, but proposing a building of the same average height, and therefore the same internal content, as the developers' brief requires. As part of the negotiated deal, the developer is of course getting something else he wants: the 'Zaha Hadid original' for whose production she was hired in the first place. This clearly sets a favourable climate of negotiation from the outset. In turn, this makes it easier for the developer to accept a breach of the 'I pay the piper so I should call the tune' rule which lies somewhere under the surface of all commercial transactions, particularly since this breach is legitimated by the claim of a simple misunderstanding ('I don't understand what they are saying') and enforced by calling on a whole network of legal rules and sanctions too ('you did tell me in writing'). Finally, hanging silently in the air in this negotiation, is the fact that 'I always made faces and frowned.' The fact that she bothers to tell us this suggests that it has some significance. I should like to believe that it shows a woman from an ethnic minority using the issues of gender and ethnicity, which must so often have proved disadvantageous, as positive assets, in a negotiation where the other actors are probably mostly men, bound nowadays by at least some degree of middle-class political correctness.

To summarise, this example has helped us to focus on a number of factors which appear to be important in the form-production battlefield. First there is the question of the power available to the various actors: access to economic or political power, or to valued knowledge or cultural capital. Second, there are the rules according to which the various actors operate in the form-production process. Third, there are the sanctions through which these rules are enforced. And finally there is the issue of initiative: who gets to set the agenda about what?

So far so good: we have developed some ideas which can help us understand what is going on in the negotiations which are central to the form-production process. In the process, however, we have been brought face-to-face with (but rather glossed over) the sheer complexity of these negotiations. Let us now consider the practical implications of this complexity in more depth.

First of all, not only are there many actors, interacting in complex ways, but also they are each addressing issues which are complex in their own right – recall the issues of climate, materials, technology, 'function', market relationships and so forth which we encountered in the last chapter. As we saw then, each of these issues – even considered separately – comprises a web of loosely-defined considerations, complexly connected into social, political, economic and cultural domains.

Clearly this is a field of work which cannot be carried out by some systematic process of generating and evaluating all the possible options for action. If we try to do so – as some did during the 1960s, for example

– we find ourselves in the dilemma identified by the American design theorist John Eberhard, in this amusing (but horribly believable) account from that time:

> This has been my experience in Washington when I had money to give away. If I gave a contract to a designer and said, 'The doorknob to my office really doesn't have much imagination, much design content. Will you design me a new doorknob?' He would say 'Yes', and after we establish a price he goes away. A week later he comes back and says 'Mr Eberhard, I've been thinking about that doorknob. First, we ought to ask ourselves whether a doorknob is the best way of opening and closing a door.' I say, 'Fine, I believe in imagination, go to it.' He comes back later and says 'You know, I've been thinking about your problem, and the only reason that we have to worry about doorknobs is that you presume you want a door to your office. Are you *sure* that a door is best way of controlling egress, exit, and privacy?' 'No, not at all.' 'Well, I want to worry about that problem.' He comes back a week later and says, 'The only reason we have to worry about the aperture problem is that you insist upon having four walls around your office. Are you sure that is the best way of organizing this space for the kind of work you are doing as a bureaucrat?' I say 'No, I'm not sure at all.' Well, this escalates until (and this has literally happened in two contracts, although not through this exact process) our physical designer comes back and he says with a very serious face, 'Mr Eberhard, we have to decide whether capitalistic democracy is the best way to organize our country before I can *possibly* attack your problem.'[25]

Lest anyone imagines that it might be possible to overcome this problem with the aid of some new generation of supercomputers – admittedly these did not exist when Eberhard wrote his story – let us remember that we should still be faced with the further level of complexity which flows from the difficulties of co-ordinating and controlling the many members of the so-called 'development team', a difficulty which deepens by the day, because the complexity of the form-production process itself appears everywhere to be increasing, though it has advanced further in some countries than in others. At its most complex, in countries like the UK and the USA, the development process involves many professionals influencing the form-generation process alongside the architect.

In discussing how the development process works, Cadman and Austin-Crowe point directly to issues of co-ordination and control:

> In order to be really effective, each of these separate roles must be combined within the development team. Indeed one of the most important functions of the developer is to be able to select and bring together a team of advisers who complement each other and work well together.[26]

And yet things somehow get done. And, more surprising still, they seem to get done more or less to the satisfaction of all the mutually ignorant and faintly hostile actors who are involved in the form-production process. We do not always find property developers making spectacularly low profits. Nor are our prisons or psychiatric hospitals full of architects who have blown up their buildings or had nervous breakdowns. Indeed, property developers sometimes make very handsome profits, and it seems that these are not necessarily achieved at the cost of unbearable angst amongst the architects involved. Quite the contrary: the evidence suggests that architects on the whole enjoy their work. In Britain, for example, they are willing to undergo a seven-year period of professional training, in order to join one of the worst-paid and least-respected professions in the country.

The compensation is a high level of job-satisfaction; and when we ask where this comes from, we find that it stems from the 'creative design' aspect of the work. Studying 600 German architects in 1965, for example, Bolte and Richter found that the statement 'my chosen profession should give me the opportunity to do creative work' was chosen as the most important of a number of alternative views by 66 per cent of the architects involved;[27] whilst Salaman – questioning 52 London architects in 1970 – found that for 63 per cent of them 'creativity plus design enjoyment' gave the major part of their work satisfaction.[28] If anything, this orientation may be strengthening. In her 1979 study of over 400 architects in 152 Manhattan firms, for example, the sociologist Judith Blau found that 'of the architects interviewed 98 per cent mentioned creativity as the distinctive feature of architecture when compared to other professions'.[29]

On the face of it, all this is hard to understand. Patrons cannot themselves design, and have difficulty in controlling the efforts of those who can. And yet, in most instances, their complex interests seem to be satisfied, at least to an extent they can live with, through the creative efforts of architects and other professional advisers who, when not actively hostile to those interests, are primarily concerned with other issues altogether. In reaching this point, we have gone as far as the various strands of the 'problematic of action' can take us. We have seen that though human action is central to the form-production process, we cannot understand that process entirely as the out-come of the actions of heroic individuals, nor as the result of orders handed down from masters to servants, nor through the co-ordinating effects of market signals. Far more convincing is the more complex understanding offered by the 'battlefield' problematic, in which actors deploy their resources of economic or political power, valued knowledge or cultural capital, in more or less adroit ways, in attempts to make things happen as they want.

Even this more sophisticated problematic, however, has only taken us so far. Eventually it has left us with an apparent paradox: it seems as though something 'above' all the various actors must be co-ordinating their actions. But in the last chapter we saw that it is not plausible to imagine that built form is determined by factors 'outside' human action. How can individual actions be co-ordinated by something which is not outside themselves? If we are to move forward, that is the question which the next chapter will have to address.

3   Genius and tradition

At the close of the last chapter, we were faced with an apparent paradox. To move forward, we have to understand what it is that co-ordinates the various actors in the complex division of labour of the modern form-production process, given on the one hand that they neither understand nor sympathise with each other's objectives, and on the other hand that co-ordination cannot simply be imposed by one actor on another, by force.

Since co-ordination cannot be imposed on any actor from outside, it must somehow be established at least partly from within: there must be some form of co-ordination going on 'inside each actor's head', so to speak. But we have also seen the importance of individual action and initiative in the form-production process, so there has to be scope for this too inside actors' heads. Is there a practical problematic which allows space both for individual actions and for some process through which these are co-ordinated? Perhaps we can move our understanding forward by drawing on the long-standing body of design theory which stresses 'on the one hand our appreciation of the imaginative genius of the individual and, on the other, the importance of architectural tradition', as the architectural historian David Watkin puts it.[1]

The flavour of what is meant by this distinction between an individual, private 'genius' and the more public, co-ordinating heritage of 'tradition' in the form-production process comes over very strongly in this account by the Swiss architect Mario Botta:

> What I think about architecture today is the result of the information and culture of [sic] which I have been exposed. To a certain extent it is a question of the cultural and collective heritage transmitted by previous generations, the whole set of ideas and thoughts that have conditioned and nourished my education. In other words, this heritage, the direct or indirect inheritance of what has previously been made, forms the theoretical condition for being an architect today.
>
> What I *think* therefore are the sufficiently rational and defensible aspects of the social and collective values that contribute our discipline. Values that can be critically and reasonably evaluated.
>
> On the other hand, the *feelings* I have towards architecture concern

the most subjective, the most autobiographical and to some extent the most secret aspects of myself. Together these constitute the irrational motivations (sometimes difficult to define) that also intervene in the elaboration, the evaluation and the selection of a project.

Everything I say is influenced by an alternation between these two poles: public and private, rational and irrational.[2]

Let us start by exploring the 'tradition' side of this 'genius and tradition' relationship – what Botta calls 'the direct or indirect inheritance of what has previously been made' – for it is this which has the potential to explain how actors' actions are co-ordinated. For many people, the concept of tradition has strong and negative connotations of backward-looking nostalgia, standing in contrast to more positive terms like 'modernity'. Thinkers over the last forty years, however, have amply shown how false it is to claim this dichotomy between tradition and modernity.[3] Across the gamut of human practices, from the sciences to the arts as much as in everyday life, it has gradually become clear that modernity and tradition are inextricably intertwined in practice.

In relation to the scientific world, for example, Brian Wynne tells us what the sociology of science suggests: 'there has never been any such thing as modernity in the mythic sense described in the rhetoric of scientists . . . Scientific 'modernity' has always been involved with tradition.'[4] Despite – even, perhaps, because of – its rhetoric of originality and innovation, this is equally true across the field of the arts. The search for originality, for example, cannot usefully be understood as an attack on tradition. Rather it has come to constitute a tradition in its own right: *the Tradition of the New*, as art theorist Harold Rosenberg calls it.[5] To acknowledge the importance of tradition in the conduct of human affairs, therefore, is not by any means to abandon a 'modern' open-minded spirit of scientific enquiry or artistic creativity. It is merely to accept that much of what goes on in these processes is guided by practical, tacit knowledge which is itself unexamined and traditional in its nature – necessarily so, since the factors which influence anyone's practical actions are always far too complex ever to be verbalised or theorised completely.

Traditional knowledge – Botta's 'heritage transmitted by previous generations' – is therefore always important in the form-production process, as in any area of practical life. But before we can begin to understand the role it might play in co-ordinating the actions of different actors, we have to understand more about what kind of entity tradition might be.

According to Botta's practical insights, tradition has two complementary characteristics. On the one hand, it does not just happen; rather it is created through human action as the 'inheritance of what has previously been made'. Once made, on the other hand, it affects new actions in some all-encompassing way; in Botta's view, it 'forms the theoretical condition for being an architect today'. Tradition, in this view, is somehow both the outcome and the medium of the 'public pole' of human involvement in the form-generation process. But this raises further questions. How is tradition 'inherited' by one generation from another? And how, in turn, does it act as a medium for new design work? These are not the sorts of questions which practitioners are often called on to answer in their everyday work. To answer them, therefore, we have to move for a while from the world of practice to that of social theory.

The ideas we need from social theory have to be consistent with the insights we have so far developed from the world of practice. Highly relevant for our purpose, in these terms, is the 'theory of structuration' developed by the British sociologist Anthony Giddens.[6] Central here is the concept of the 'duality of structure', according to which structural properties, which give social systems their continuity across time and space, are seen both as the medium through which social action takes place, and as the outcome of that action itself. Clearly, Botta's suggestions about tradition fit very directly into this framework: tradition, as we have seen, is itself both the medium and the outcome of social action in this sense. Giddens's theory therefore offers us a conceptual framework which we can use to set what we have already discovered into a wider context. It can also help us to go beyond this point, to learn more about the formation of the complex set of structures we call 'tradition', through the insights it offers about the nature of structures in general.

In common usage, and in particular to many likely readers of this book, the word 'structure' has connotations of solid permanent things such as columns and girders. This is not at all what it means for us here. Social systems do not have static 'structures' in this sense, in some way separate from human action going on around them. Rather, the structural properties which give them continuity through time and across space are embodied in what Giddens calls the 'rules and resources which are implicated in the reproduction of social systems'.[7] Let us explore these rules and resources in more depth.

The rules of social life, says Giddens, should be regarded as 'techniques or generalizable procedures applied in the enactment/reproduction of social practices',[8] whereas resources are 'the means material and symbolic, whereby actors make things happen'.[9] Though it is useful to separate them for analytical purposes, it is important to realise that rules and resources are indissolubly linked in practice. Practically speaking, a rule is only a rule if it is supported by sanctions against breaking it; and sanctions depend for their effects on the mobilisation of resources. The resources which can be mobilised to make things happen come in various forms, as we saw in the last chapter. At the material level, for example, they include economic and political power, as well as physical structures which people can use to support a particular course of action. Symbolic means include 'knowledge power' as well as the ability to confer or to withhold whatever counts as prestige in any given social milieu – 'cultural capital', for example.

In terms of the rules themselves, Giddens suggests that 'all social rules have constitutive and regulative (sanctioning) aspects to them'.[10] In un-ravelling how rules work, it is useful to explore this analytical distinction a little further.

The 'constitutive' aspect of social rules is concerned with defining what exists and what does not. That is, as the Swedish sociologist Goran Therborn puts it, 'who we are, what the world is, what nature, society, men and women are like. In this way we acquire a sense of identity, becoming conscious of what is real and true; the visibility of the world is thereby structured by the distribution of spotlights, shadows and dark-ness.'[11] The constitutive aspect of rules enables us to get to grips with the otherwise impossibly complex flux of the world in which we live; without them, we could not hope to survive, let alone take effective action in a complex process of form-production. As well as enabling, however, they also constrain us in important ways, because definitions of what exists set boundaries to the concepts we use: 'this is an x, that is a y.' Like the walls of a building, which affect where we can go and where we cannot, but even more fundamental in their impact, the conceptual walls erected at the boundaries of definitions affect the kinds of thoughts it is possible for us to think.

The 'regulative' aspects of rules are more complex, because they operate at more than one level. In one sense, they are about evaluation: in Therborn's words, '*what is good*ç, right, just, beautiful, attractive, enjoyable, and its opposites. In this way our desires become structured and norm-alized'.[12] By itself, though, this is not enough to 'regulate' social action. It is also necessary to define *what is possible* and what is not; at another level, the regulative aspects of rules are about this too. They are about 'our sense of the mutability of our being-in-the-world' and, says Therborn, how the 'consequences of change are hereby patterned, and our hopes, ambitions and fears given shape'.[13]

The rules and resources which keep social systems going are partly external to particular agents, embodied for example in legal systems, economic practices and structures of built form. In addition, however, rules and resources are also embodied in 'the memory traces which orientate the conduct of knowledgeable agents',[14] as Giddens puts it. Structures are 'in people's heads' as much as 'out there'. Roark's *Fountainhead* rule about originality, for example (see p. 29), is partly in his own head. So is the resource – in this case, his own commitment to a vision of himself as an artist – through which transgressions of that rule are prohibited by inducing feelings of guilt. On the other hand, insofar as the 'be original' rule is more widely shared amongst his peers, there are also external sanctions: the shame or loss of social honour which (in that circle) would follow from being seen as 'unoriginal'. But again, other rules and resources are totally external in Roark's case. Various rules of economic action, for example, are entirely foreign to him. The sanctions through which these rules are enforced are also applied entirely from the outside: external pressures of bankruptcy and imprisonment, rather than internal feelings of guilt.

Once actors have taken their vocabularies of rules on board, these are called on to play a key role in the conduct of their everyday lives. But how are these rules taken on board in the first place? What drives the learning process through which they get into actors' heads? Because of the funda-mental role which rules play in the process of human development, this learning process has to begin from a very early age. Each new human infant, with manifold potentials for individual development, is born not into a vacuum but into a social situation which is already structured according to a complex set of rules and resources which are 'given', so far as the infant is concerned.

Right from the beginning, the infant's experience of the world is beset with inherent ambiguities. The psychoanalyst Melanie Klein suggests, for example, that even the earliest experience of the mother's breast is highly ambivalent: it is the major source of satisfaction, but because it is not always available when wanted, it is the major source of frustration too.[15] The fact that two opposed feelings are associated with the same object – a situation which recurs throughout human life – does not make sense of its own accord. But the need to make sense of life in this inherently contradictory world is fundamental to the human condition, for a senseless world is not a viable one in psychic terms. In such terms, indeed, senselessness may be worse even than suffering, as Nietzsche long ago suggested: 'Man, the bravest of animals and the one most accustomed to suffering, does not *repudiate* suffering as such; he *desires* it, he even seeks it out, provided he is shown a *meaning* for it . . .'[16] Nearer our own time, the psychologist Rom Harré confirms the psychic importance of sense-making when he claims that '[the] ultimate problem for a human being is to make his actions intelligible to himself, so as to preserve his sense of personal worth'.[17] The idea that this need to make sense also has important implications for the structural properties of social systems is voiced by Cornelius Castoriadis when he writes of society being founded on a belief in 'its claim to render the world and life coherent'.[18]

Sense is given to life through the internal rules and resources which constitute that aspect of each actor's personality which we call 'subjectivity'. Subjectivities are the learned parts of actors' overall personalities: 'the forms of human subjectivity are constituted by the intersections of the psychic and the social, and may be seen as the outer, more conscious and more socially changeable aspects of the person', Therborn suggests.[19] Because experience of the 'external' world does not automatically make sense, it follows that the process through which subjectivities are developed cannot merely be a passive 'absorption' of the structural regime into which one is born. Sense has rather to be made actively, through the use of the human potentials with which each infant is endowed: 'the creative working of unconscious desire', working through imagination, as Anthony Elliott puts it.[20] Here is a nice example of this active process in practice:

> A small toddler, just learning to talk, sat on his father's lap while their plane taxied down the runway in preparation for take-off. The two were looking out the window at planes moving along the runway behind them. 'Airplane', the father says. 'Car', responds the child. 'No, it's an airplane', corrects the father. Undaunted, because he *knows* that airplanes fly in the air and vehicles move on the ground, the child repeats, 'car'.[21]

The active process of subjectivity-construction has a dual character. On the one hand, it is a process of coming subjectively to terms with the objective, external framework of rules and resources which demarcate one's life-chances, a process of 'matching' or 'fitting in', so that life makes sense because one comes to *want* to do what one *has* to do. As Therborn puts it, 'human infants are subjected to a particular order that allows or favours certain drives and capacities, and prohibits or disfavours others.'[22] On the other hand, the process of subjectivity-construction is enabling as well as constraining in its nature, as Therborn goes on to point out: 'At the same time, through the same process, new members become qualified to take up and perform (a particular part of) the repertoire of roles given in the society into which they are born, including the role of possible agents of social change.'[23]

Sometimes these enabling and constraining rules are codified and written down, as in planning briefs or building regulations. Many, however, are not; including some of those which are most important in the reproduction of social systems. Partly this is because they do not need to be codified, precisely because they are so deeply embedded in the conduct of everyday affairs. More fundamentally, however, it is because the rules which we call on in running our everyday lives are so complex that they would be impossibly cumbersome to use at a conscious level. Think, for example, of the 'simple' process of catching a ball, and consider the complex rules governing bodily movement, and the knowledge-resources about trajectories, air resistance and so on which must be called on for its success. The moment we try to use these rules and resources consciously, the ball has already passed us. Given that the performance of everyday life overall is infinitely more complex than this relatively simple routine, it would be impossible in practice to run our lives at all if we had to rely only on conscious rules. Though they include such processes as speech and writing, in which consciousness plays a large part, the 'techniques and generalisable procedures' in which rules are embodied are very often well below the threshold of consciousness.

From our point of view, it is important to recognise that techniques such as speech and writing are in any case not very good for embodying rules about built form, since forms are themselves notoriously difficult to define in terms of words alone: 'There is no way to perform architecture in a book' as Bernard Tschumi pithily reminds us.[24] Given this problem, how are rules about built form embodied?

Here we can return to the world of practical knowledge to help us once again, through the longstanding school of architectural thought which holds that such rules are embodied through the medium of 'types'. As architectural theorists Karen Franck and Lynda Schneekloth see it, for example,

> Humans do not occupy, imagine, or create an infinite variety of particular, idiosyncratic places. Instead, we structure environments by creating and using a multitude of categories of places and spaces, often called 'types'. With these types we group places that are alike together and we treat individual places as members of groups. This ordering of space into different kinds of spaces is an intrinsic and constituent part of life.[25]

The repertoire of types in good currency in any particular social context forms a frame of reference which enables people to make sense of occupying, imagining and creating their worlds, by breaking down the overwhelming variety and complexity of those worlds and their potentials into simpler, more graspable structures. The typological repertoire as a whole makes this possible because it contains 'typological chains' which have a hierarchical character, linking the 'surface' complexity of the world with a far simpler set of 'deep' types (Figure 3.1).

From the form-production perspective, both 'deep' types and those nearer the 'surface' of production culture each embody the irreducible minimum specification for generating some formal or spatial element, or some relationship between these, which is part of the material dimension of a given culture or subculture — what it is to be a street or a park, for example. All the particular forms which people generate by using any particular type will be

	SQUARES	LAMP POSTS
	which are made up from...	which are made up from...
PUBLIC SPACES	which are made up from... STREETS	which are made up from... SIDEWALKS
	which are made up from...	which are made up from...
	PARKS	BOLLARDS

DEEP TYPES   **SURFACE TYPES**

Figure 3.1 Deep types and surface types.

constrained by this specification; but they may all be different in all other respects. Types, therefore, are not in themselves designs. Rather they each constitute the cultural raw material from which a large 'family' of different designs can be generated. As the French architectural theorist Quatremère de Quincy put it long ago, 'the type . . . is an object with respect to which each artist can conceive works of art that may have no resemblance to each other.'[26] An artificially simple example, proposed by Hillier and Leaman, may help to clarify this distinction between type and design:

> An army marches all day. At nightfall a halt is called and unpacking begins. Within a short time a structured environment appears. Tents of various kinds and sizes are placed in certain definite relations; kitchens, sentry posts, flags, fences and other paraphernalia are erected. A complete environment is, as it were, 'unfolded'.[27]

Here the deep type evoked in English by the word 'camp', but probably common to any army on the march, is linked to the potentially infinite variety of actual camps which might be set up through a chain of types nearer to the surface of any particular army's culture. These relatively 'shallow' types, still generic in nature, stay the same from one night's camp to the next, defined in words in a set of instructions in the army manual, and tacitly in the prefabricated bits and pieces which are used to construct each camp. But the particular camps which are constructed after each day's march are all likely to be different, because of differences in terrain, changes in the number of soldiers to be accommodated and so forth – all of which have to be taken into account when those responsible for putting up the camp interpret the type-manual.

At both deep and shallow levels, the rules about form which are embodied in types have both constitutive and regulative dimensions. In the constitutive realm, types play a central role in defining 'what the world is': 'this is a street and that is a square, and over there is a park', for example. Taken together, the vocabulary of types in good currency at any particular time and place also regulates the limits of what is possible, or even think-able. According to Franck and Schneekloth, for example, 'typing enables us to create order in the world and to make sense of our lives as individuals and as groups of people . . . we cannot live in ways we cannot order, or in ways we have not yet ordered.'[28] In this view, types – embodying both consti-tutive and regulative rules – must play a crucial role in co-ordinating the form-generation process. How credible is this idea?

First, it gains support from the ways in which certain well-known architects explain their own processes of design. The distinction between a 'type' and the particular design which is generated from it is surely the same, for example, as the distinction which Louis Kahn was fond of making between 'Form' and 'Design', where 'Form' has clear connotations of 'generic type'. As Kahn himself put it,

Form is what, Design is how. Form is impersonal, but design belongs to the designer. Design is prescribed by circumstances – how much money there is available, the site, the client, the extent of skill and knowledge, and above all the individual's tendencies of expression.[29]

'Form', in contrast, has no such particular characteristics, and is not created by the individual designer: in Kahn's words, 'The form "symphony" does not belong to the composer. His design, his composition does.'[30] Indeed, if we accept that no design is created *ex nihilo*, then the only alternative to drawing on some repertoire of types is to crib directly from someone else's design. This is clear, for example, to Igor Stravinsky – surely himself one of the twentieth century's most creative artists – when he insists that 'everything that is not tradition is plagiarism.'[31]

Just because certain creative people acknowledge the role of types in their own work, however, does not mean that all acknowledge this – indeed, most do not. Perhaps the use of types in design is only one particular approach, which is taken only by a particular clique; is there any evidence that the role of types in design is more universal than that? The idea of a set of 'types' embedded in culture, and therefore experienced as 'given' by individual designers, is indeed supported by intuitions from many sources. It makes sense, for example, of the otherwise very odd way we talk about the non-professional design processes which make up so much of our everyday lives. Here, in any kind of problem-solving activity, we say (and we say it within many cultures, in many languages) that we have 'found' a solution ('Eureka!'), rather than (say) 'invented' it; as though the solution – in some general form, at least – was somehow already available to be found.

The idea that the 'recognition' of pre-existing types has a central role in design is also supported very clearly by a number of more formal investigations of the architectural design process. The architect and psychologist Bryan Lawson, for example,[32] argues that researchers in design psychology seem broadly to agree that the stages which designers typically go through in the process of design are centred around a moment of 'illumination', sometimes referred to as the 'Eureka phenomenon' after Archimedes's celebrated cry of bathtime enlightenment.

If professional designers use types when they design, this approach is not something they learn through specialised design education, for children (for example) do it too. Young children have not had either the time or (mostly) the opportunity to become strongly influenced by the formal cultures of design; and in consequence, as the psychologist Edward de Bono puts it, their designs 'achieve an effect that rises above petty rules of procedure'.[33] And yet, when we analyse their designs – even designs for something which itself has no precedent – we find that types play an important role.

Analysing children's designs for 'a machine for exercising dogs' (Figure 3.2), de Bono felt that these lie 'somewhere between poetry and reason and humour and art' – wonderfully free and creative efforts. In analysing how

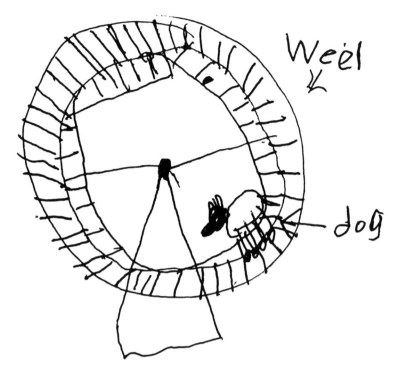

Weel

dog

Figure 3.2 A Machine for Exercising Dogs, by Jonathan Williams.

the children handled the process of design, however, he finds himself emphasising the role of precedent: 'To a child most ideas are new ideas. And yet to arrive at these new ideas he can only make use of the ideas he already has.'[34] But in using these 'ideas he already has' the child cannot simply be copying precedent, since no child has ever before seen a machine for exercising dogs. If they are to be useful in creating something new, the ideas have to be in an abstracted, generalised form. The machine for exercising dogs has to be creatively conceived as the same type of thing as something the child has already experienced: a type of robot, a type of treadmill, a type of seesaw or whatever. In other words, even inexperienced designers with no formal training use types in design.

Further support for the idea that types are fundamental to the process of design is given by evidence that they are called on, in practice, even by designers who have a strong desire to stress their own originality. For example, Frank Lloyd Wright, who placed an almost Howard Roarkish value on personal creativity, and emphasised that aspect of his work throughout his own writings (he had a penchant for titles like *Genius and the Mobocracy*)[35] nevertheless used a relatively small number of generic types in his own massive output of highly individualistic design work. March and

Steadman, for example, show very convincingly that a number of Wright's houses, including some which appear at first sight to differ markedly in both visual and spatial terms, are all generated from the same typological structure (Figure 3.3).[36] The type (of spatial structure in this case) is the same in each situation, but the various designs each have their own unique characters.

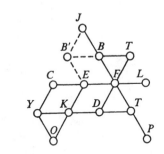

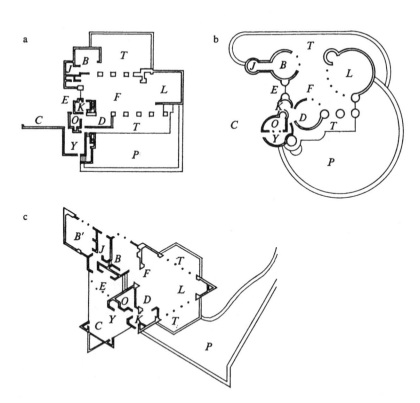

Figure 3.3 Different plans but the same spatial type: houses by Frank Lloyd Wright.

Overall, then, there are powerful reasons for accepting some sort of 'type plus creativity' approach in explaining how the form-generation process works. In practice, however, this sort of explanation is certainly not as well accepted, by the general run of design professionals, as the approaches we have reviewed in the last two chapters. As the architectural theorist Micha Bandini points out, 'typology was probably only a major issue for architecture during the 1960s and 1970s.'[37] Following the logic of our own argument, therefore, we have still to be cautious about using this approach as the basis for developing our own explanation of the form-generation process. Specifically, we have to explain why it is not in wider currency: why is it not the most widespread explanation? Is this for reasons which should make us reject it ourselves?

One obvious and important reason is that there is an apparent tension between the concepts of originality and creativity, so dear to modern design professionals, and the role of types in the design process: since the types which designers use must already exist before any particular situation where they are used, they cannot themselves be seen as the product of the individual designer's creativity; and this directly contradicts the 'Fountainhead Myth', so tenaciously held by so many.

This contradiction makes many modern designers and design theorists see the 'idea of type' as a problem, rather than merely as a central aspect of how design is possible. For example, in referring to the work of the 1930s Gestalt psychologists Wertheimer, Duncker and Maier, Bryan Lawson points out that these researchers found that 'each problem is not viewed afresh but rather is first classified according to types of problems already encountered, and the solution selected accordingly.'[38] Given the role of types in our emerging explanation of the form-generation process, this is exactly what we should expect to find. Lawson, however, refers to this in the most pejorative terms, as 'the almost universal tendency of experience to have a mechanising effect on individual thinking', and he goes on to discuss a range of techniques for trying to avoid it.

It is not too difficult to understand the ideological dimension of attitudes like this, or of the relative silence about types in the recent design theory literature. We have already seen that it is important for designers to claim a strong degree of originality in their work, if they are to compete in the market for design services. Clearly this does not fit in well with the idea that designers are 'merely' selecting and manipulating socially-sanctioned types. From this point of view, a hostile silence is only to be expected. This is yet another case where the practical value of a problematic – this time through the 'silent discourse' about what it does *not* express – lies precisely in its ability to hide the workings of the form-production process from those who are engaged in it. When we understand this, we can see that the silent discourse is far from destroying the credibility of the 'type' approach, from our point of view.

In any case, worries about the 'type' approach leaving no space for individual genius are wide of the mark. The repertoire of types in good currency, forming part of any particular cultural situation, is not a fixed one. It changes over time, and if we are interested in how it changes, we have to focus on individual human action, moving from a focus on 'tradition' to the complementary role of 'genius': 'the most secret side of myself', as Mario Botta put it. How is this 'private' pole of the form-generation process formed? Both 'nature' and 'nurture' components are involved here. On the 'nature' side, it is important to remember that no agent's personality can ever be fully described through its subjectivity. There always remains that part of the overall personality which cannot be accounted for as a result of subjective development, but which has rather to be seen as one of its driving forces: the biochemically fuelled 'desiring machine', to use the unforgettable phrase coined by Gilles Deleuze and Felix Guattari,[39] which gets the process of subjectivity-construction off the launchpad around birth, and keeps it in orbit till death at least. Currently, the most widespread conception of desire is more passive than this. As Deleuze and Guattari point out, desire is commonly thought of as connected with the idea of acquisition, which 'causes us to look upon it primarily as a lack'.[40] I shall give the name of 'aspiration' to this lack-orientated, more socially-directed outcome of desire, which takes particular directions under particular social contexts, to distinguish it from the more open, exploratory character of desire itself, as described by Elisabeth Grosz: 'Desire does not take for itself a particular object whose attainment it requires; rather, it aims at nothing above its own proliferation or self-expression . . . It moves, it does.'[41]

On the 'nurture' side, the creative workings of unconscious desire bang into and interact with the 'givens' of the social milieu to generate subjectivities whose rules and resources are linked to characteristics of the milieu itself. It is important to realise, however, that this is far from implying any neat comprehensive 'fit' between the individual agent and the milieu concerned. At least in complex modern societies, the subjective development of any particular agent, taking place in relation to such varied institutions as family, gender, school, race and so forth, is itself complex, to the extent that any particular person must effectively develop multiple subjectivities, of different types; as Therborn explains:

> A single human being may act as an almost unlimited number of subjects, and in the course of a single human life a large number of subjectivities are in fact acted out. In any situation, particularly in a complex modern society, a given human being usually has several subjectivities that might be applied . . . For example, when a strike is called, a worker may be addressed as a member of the working class, as a union member, as a mate of his fellow workers, as the long-faithful employee of a good employer, as a father or mother, as an honest worker, as a good citizen, as a Communist or an anti-communist, as a Catholic, and so on.[42]

As they develop their multiple subjectivities, individuals acquire increasing abilities to negotiate their own ways through the various situations with which they are confronted. The extent to which this ability develops in practice depends on all sorts of factors which are particular to individuals. Some of these have, no doubt, biological components, to do with issues of intelligence-potential, physical strength and so forth. Others are completely social, involving accidents of birth in terms of parental culture, economic standing and the like. As chaos theory suggests, even apparently small and insignificant early variations in such factors may contribute to radically different eventual outcomes. Some people will negotiate themselves through this matrix into an involvement with the production of urban form. Everyone save the hermit, whether they intend to or not, will end up involved in using it. Given the widespread dislike which many users seem to feel for many recent transformations, there is always the potential for conflict between the 'user' and 'producer' sides of the subjectivities of people involved in the form-production process. To maintain the process of making sense of their working lives, those involved in producing built form are drawn to ideological material which helps them to keep these conflicting aspects of their personalities apart, but this can never be entirely successful: since they are embodied in the memory traces of the same biological individual, there is always some likelihood of 'leakage' between one subjectivity and another.

This leakage, and the conflicts to which it gives rise, always offers some potential for loosening any particular producer's commitment to maintaining the status quo so far as the form-production process is concerned. The interplay between multiple subjectivities, and between subjectivities and Elliott's 'creative workings of unconscious desire'[43] therefore means that types, though themselves inter-subjectively shared, are called on by different actors in all sorts of different ways. It seems, however, that even those designers with the most nonconformist personalities – designers who, like Le Corbusier, rebel against earlier design traditions in the most self-conscious way – still have to call on established types in their work, as constitutive 'raw material' to which new regulative rules can be applied. As the architectural historian Peter Collins points out, for example, many of Le Corbusier's most innovative ideas were generated by inverting or reversing existing types:

> when compared with traditional building methods, they constituted a kind of 'anti-architecture'. For example, whereas in traditional architecture, a villa is situated in a garden, in Le Corbusier's architecture, the garden is situated in the villa. Whereas in Classical architecture colonnades are placed on a base of solid walling, Le Corbusier places solid walling on top of his columns.[44]

Even the buildings of Le Corbusier's last period, after the Second World War – which in so many other respects mark a radical departure from his earlier works – continue to show this characteristic. The Unité d'Habitation at

Marseilles, for example, shows all sorts of further reversals of traditional types in housing. Here the roof is treated as if it were the ground in a traditional scheme, with running track, terrace and other outdoor facilities, interspersed with large natural rocks. The interior too inverts traditional types: the street, which in a traditional scheme would have been outside the buildings at ground level, is here placed inside and half-way up the building. Types are used here, but they are used critically and creatively (words which must not necessarily be taken as implying admiration for the results). They are used as springboards for design, rather than as straitjackets into which the design process is forced.

Le Corbusier, of course, was not alone in this approach. In general the evidence of the avant-garde suggests that, in practice, the hunt for innovative forms always has to proceed through the medium of established types, either by using them 'reversed', as in the examples above, or by importing them from outside established architectural culture: for example, in Le Corbusier's case, from sources such as grain silos or ocean liners. Whether designers self-consciously try to be innovative (like Le Corbusier and many others of the modernist avant-garde), or whether they address design situations in less self-conscious ways (like Edward de Bono's children), it seems clear that particular designs are generated through the interaction between the multiple subjectivities of particular designers, together with 'the most secret aspects of themselves', calling on the range of types which is culturally available to them.

Whether innovation is intended or not, this process always generates innovations in practice, because each design situation is itself necessarily unique, at least to some extent. Actors in the development process who see any particular innovation as offering them advantages, in terms of their own rules and resources, will deploy whatever resources they have in support of the innovation concerned, in preference to any more established alternatives which might be available. Different actors within the development team, however, will often see quite different kinds of advantages or disadvantages in any particular innovation. To one actor, for example, the innovation concerned might seem to offer economic advantages, whilst another might see only disadvantages in (for example) social or aesthetic terms. This means that there are constant disagreements, within the development team, about whether any particular innovation should be seen in positive terms and therefore replicated in their own work, or in negative terms and therefore rejected. Given the complex nature of the work which all members of the team carry out, and the consequent difficulty of controlling their actions in detail, these disagreements cannot usually be resolved by one party merely giving orders to the others. As we saw in the last chapter, attempts to resolve them have to depend on complex (often unspoken) processes of negotiation, whose outcome depends on the particular tactics through which the various parties deploy their different kinds of resources, and on the alliances which they form between them.

Even when the upshot of this negotiation process is a decision to replicate some particular innovation, however, it is in practice impossible to transpose particular innovative designs directly for use in other situations: they are too concrete and individual in character to be able to 'travel' elsewhere. All that can be taken from them are certain aspects or qualities which they possess, so the transposition process involves abstracting these qualities from the particular, selected designs. The qualities to be abstracted, of course, must be the qualities for which these designs were chosen in the first place: these are the qualities which are 'accepted' across the agents in the development team. In other words, the value of the selected design, to the development team as a whole, resides in these qualities. But the design does not come bearing labels defining the relevant qualities. How, then, are these particular qualities abstracted from the welter of other, irrelevant qualities which all real designs must also possess? And how are these abstractions absorbed into design culture, to become 'available' as new types for other designers to use?

The first step in the process of abstraction occurs when the advantageous qualities are 'recognised' by agents in the development process, as being worth the hassle of departing from previously accepted ways of doing things. This 'recognition' may be verbalised and debated amongst the various members of the development team. Equally, though, it may take the form of a tacit recognition of the new idea's advantages, at a 'gut' level – recognised because the actors concerned are already sensitised to the typical problems they face, and can spot potential solutions, when offered, a mile off.

The next step in the abstraction and dissemination process often takes place through the specialised professional journals relevant to the various members of the development team; which are always on the lookout for news in the form of innovative projects. Here again, the particular qualities which make the project advantageous may be spelled out intellectually, or the project may be published only as a graphic image, from which the journal's readers – or such of them as face similar problems in their own work – can themselves 'recognise' the innovation's advantageous qualities. By this stage, now available to a wider professional audience, the original project has become a 'model' which other designers can use in their own work.

If the original model is adopted and adapted by many designers, it becomes clear through practice whether its qualities can advantageously be reproduced in different situations. In the course of this experimentation, both the qualities themselves and the 'rules' about when they can and cannot be achieved, become gradually clearer. In parallel, seeking to provide useful material for their readers to boost their own circulation, the professional media (capitalist enterprises themselves) produce intellectualised discussions of the physical design principles which underlie the advantages of the new models. Once the process of abstraction has reached the point where it is these principles themselves which form the focus of interest in magazines and books, with the individual models being conceived merely as particular manifestations or instances of these principles, which cannot be further abstracted without losing their point, then the various models have been abstracted into a single generic 'type'.

Once this has happened, the type concerned becomes one of the assumptions around which design discourse is organised. Often, once this stage is reached, types become deeply embedded in all sorts of legislation and knowledge-systems. Traffic engineering advice, for example, is predicated around a particular type of street layout, or building regulations begin to entail a particular type of relationship between windows and walls. The working methods of the various members of the development team also become aligned around this repertoire of types: we learn how to calculate financial valuations or engineering structures for those particular types, and then we want the economic and psychological advantages of reusing that knowledge in other situations, rather than having to start again on the learning curve for a new way of doing things. In the end, the new type concerned becomes part of the unspoken structure of assumptions which we use as constitutive rules to help us make sense of the built world and of the operations we perform on it. The type has, at least until further notice, become embedded in the mainstream of development culture.

Through this process of embedding, sedimented over long timespans without serious challenge from alternative innovations, certain form-types become institutionalised as 'deep' types within the overall social system of which they form a part. Embedded in and supported by legal institutions, for example, they become extremely resistant to change. This means that deep form-types may survive even the most radical of changes in the economic or political spheres: rather than being eliminated from the form-production process, they may merely be called on in new ways by those with the power to produce built form in any new system. In one regime, for example, the deep type 'public space' might be called on to generate surface types such as boulevards or parade grounds, to emphasise the power of the state. A new regime, in contrast, might call on the same deep type to produce traffic arteries and car parks. Across the change from one system to another, despite changes in shallow types, the deep type 'public space' persists. And each time a 'deep' type is reused in another project, it becomes institutionalised still further, so that eventually it may come to seem like a fundamental element which is 'inevitably' involved in the constitution of any city – a 'morphological element' rather like an element in chemistry. In practice, therefore, change can never take place at an equal rate across the entire typological repertoire: 'deep' types of morphological elements might last unchallenged for hundreds or thousands of years, whilst the more ephemeral 'surface' types can change far more quickly.

As well as understanding how any given vocabulary of types in good currency is involved in the form-production process, we can now see how that process itself contributes to the institutionalisation of morphological elements, at the same time as it continually generates new types nearer the cultural surface. All in all, then, this exploration from the structurationist viewpoint has yielded many useful insights into how the form-production process works.

To begin with, it has allowed us to synthesise the apparently contra-dictory insights of the influential but contrasting approaches we reviewed in the first two chapters. As we should expect from the widespread respect they command from practitioners, neither of these approaches has turned out to be entirely wrong. Looking from the structurationist perspective, however, we can now see that they both fall into the error of mistaking an important partial truth for the whole story. On the one hand it is true that structures with names like 'function', for example, are central to the form-production process – as the 'medium of social action', the process could not work at all without them. On the other hand, we see that these structures do not originate 'out there' beyond human action. 'Function', for example, is the word we use to sum up the complex set of rules and resources which constitutes and regulates the relationships between built form and the pattern of human activities in any particular social milieu. And further, we also see that these rules and resources are themselves constructed and constantly reconstructed through creative human action. They are the out-come of social action, as well as the medium which makes social action possible in the first place; and they could be neither reproduced nor changed over time, as they are, without creative human action: the partial truth of the 'heroic individual' finds its place here too.

Exploring the form-production process in more depth from this structurationist perspective, we have seen the central importance of the vocabulary of morphological elements and surface types in good currency, which forms the basic agenda of constitutive rules around which the power-plays of the process are articulated. This typological vocabulary is embodied through memory traces in the subjectivities of those involved in the form-production process. Because all actors in today's 'advanced' societies have multiple subjectivities, and because each actor's overall personality also involves the 'workings of unconscious desire', there is always the potential for any particular individual to make built-form proposals which call on the vocabulary of types in good currency in new and unexpected ways.

The conversion of mere proposals into real places on the ground, however, requires the agreement of those who have the resources which are needed to build, in any particular social milieu. If this acquiescence cannot be gained, then the proposal can only be still-born – it will never get beyond the stage of lines on a sheet of paper or images on a VDU. But those with the resources to build have their own subjectivities, and if they do agree to implement innovative proposals, the rules which govern their agreement may have nothing in common with those which prompted the innovations in the first place. The rules which regulate the actions of those who have the resources to build, therefore, often have a disproportionate impact in negotiations over which innovations, if any, are to be implemented.

Any innovation which passes the test of these negotiations, and is implemented on the ground, will then be further tested through practical experience. Those innovations which turn out in practice to give improved support for the interests of those with the resources to build will be desired by others in the same position, and this will support its replication on other sites, through a process of adaptation and abstraction. To survive in the market, more and more designers will find themselves under pressure to take the innovation on board, by incorporating it into their own systems of rules as 'good design', and by acclaiming its original proposers as visionaries. This means that 'what's truly visionary can be perceived only after it has been incidentally achieved', as Robert Venturi reminds us from his personal experience: 'Visionaries are visionaries only in hindsight. The only way to be visionary is to have been visionary. And – often – those who are later considered visionary were earlier considered nerds.'[45] Through this process, new visionaries will constantly add new types to the repertoire already in good currency. Eventually these will themselves be called on by new visionaries, as the cultural raw material for new innovations, in a continual process of typological transformation.

Overall, then, the structurationist approach has enabled us to find a way of combining the useful insights of the approaches we explored through 'Form untouched by human hand' and through 'Heroes and servants, markets and battlefields', without backing us into the impasses towards which these earlier approaches both lead. If this is right, we should be able to call on the structurationist approach to develop a useful explanation of the courses which recent urban transformations have followed. Before we try to do this, however, let us consolidate the position we have reached so far.

Conclusion: a framework of questions

In the last three chapters, we have worked through a wide range of approaches to explaining why urban transformations follow particular courses in particular situations. In this conclusion, we have to pull together a useful framework of ideas from the insights which these explorations have revealed.

In Chapter 1, we came to understand that we cannot arrive at any useful explanation by focusing only on disembodied, extra-human factors such as technology, even if these are considered as socially constructed. The problem here is that such an approach necessarily obscures the ways in which real human beings are involved in the processes of urban change – an approach which is negative for two reasons. First, at an intellectual level, it is unconvincing because it regards people as passive cultural dopes, ignoring the active role which human desire plays in driving any process of change. Second, at a political level, it is disempowering because it obscures any possibilities there may be for active intervention to change things for the better. Moving on through Chapter 2, however, we saw that these intellectual and political problems cannot be overcome simply by moving to the opposite strategy: we cannot get much further if we try to explain the course of urban transformations solely through the actions of particular individuals – whether singly, or in master and servant relationships, or even struggling with one another in some battlefield of change. To be sure, the more sophisticated explanations in this genre do have space for considering both individual action and the sources of power through which individuals' desires can be pursued; but they say nothing (and never can) about why the results of those individual actions on the ground are so predictable. If it is all down to individual action, why – for example – does everywhere seem to be getting more and more like everywhere else? In the end, the idea that there must be some source of structure, external to any individual, somehow influencing all individuals' actions, seems indispensable after all.

By the close of the first two chapters, then, we had seen that it is not enough to focus either on external structures or on individual actions alone. What is needed, rather, is a focus on the relationships between structure and action. In Chapter 3, we followed this way forward, drawing on the structuration approach which comprehends structures as rules and resources which have to be seen as both the medium and the outcome of people's actions. It is these socially-constructed and shared rules and resources – important amongst them form-types and canons of 'good design' – which structure individual actors' desires, so that their actions produce urban changes which exhibit a certain regularity of patterning at particular times and places.

Realising the important role which these types and canons play in the form-production process as a whole, we followed through the process of their own production in some detail. We saw that the particular actors involved in each form-production project have to call on the repertoire of types and canons which is in good currency for them; but we also saw that because each project necessarily has its own unique aspects, these structural rules always have to be interpreted in new ways, to some extent at least. This means that some degree of innovation is always called for in this form-production process, but whether any particular innovation is taken up by other actors working on other projects, to be replicated more widely, depends crucially on whether it is perceived as having advantages, in terms of their own desires, by those who have the power to choose it over any alternatives which might be available.

Different actors in the development process will often perceive different kinds of advantages and disadvantages in relation to a given innovation; for example in economic terms, or in terms of helping them to make sense of the work they have to do. We saw that the typification process is therefore not some automatic machine, but is rather itself a process of struggle: a cultural struggle whose outcome depends on the particular strategies and tactics deployed by the parties involved, and on the alliances they develop between them. Innovations whose protagonists win out in this cultural battleground will be chosen more often than any alternatives which might be available, and will therefore eventually become abstracted – through the process which we termed 'typification' – for use in a wide range of situations; and this process will generate new types to be added to the repertoire in good currency. The change to this repertoire will then have to be made sense of, at an ideological level, by those who are constrained to call on the new types in their work; we explored the ideological processes through which the forms generated from new types come to be seen as constituting 'good design', leading to changes in the current canon.

In the course of Part I as a whole, then, we have raised a framework of questions which have to be addressed if we want to understand why types and canons of good design are transformed in particular ways in particular times and places. First, we have to ask who are the key actors in the situations with which we are concerned: who has the power to initiate proposals for transformations, and who has the power to decide whether to implement them or not? Second, once this is established, we have to enquire into the rules which structure these significant actors' desires: what do they seek from the transformation processes concerned? Third, we have to ask which aspects of these transformations of types and canons offer advantages in terms of these actors' desires: it is these aspects which will have gained the support of powerful actors; and it is these aspects which will prove resistant to change, unless the changes proposed offer new advantages in their turn. These, then, are the questions we shall address in Part II.

Part II

Spatial transformations and their cultural supports

Introduction

In Part I, we saw that the form-production process takes place through a complex pattern of negotiation and struggle between various actors. We saw that the built outcome of this process depends on the internal and external economic, political and cultural resources available to each actor, and on the rules according to which the various actors deploy the resources they have. In capitalist situations, increasingly prevalent across the globe, private-sector profit-orientated development agencies deploy very large resources in the form-production process. For example, the vast majority of developments are produced by private-sector developers specifically to be sold in the marketplace at a profit. Perhaps less obviously, the private sector produces a great deal of public space too. The streets within private sector housing estates, for example, are mostly produced as integral parts of profit-orientated overall schemes, and then merely handed over to state agencies for management afterwards.

Once any private-sector project has been sold, profit-orientated economic considerations become as crucial to the buyers as they were to the original producers. This is perhaps most obvious in the case of commercial and industrial property, which is mostly bought specifically for investment, by large financial institutions such as pension funds and insurance companies, who then rent out their purchases for use by others. In turn, these end-users themselves see the properties they rent primarily in economic terms, as devices to help the cost-effective operation of their own businesses. Less obviously, but more pervasively, all real-estate properties have particular generic characteristics – not shared with other commodities – which make it very likely that all who buy them will have their future economic performance in mind, at least to some extent. First, they last for a very long time: in Britain, for example, it is estimated that only about 1 per cent of the building stock is replaced each year, which implies that buildings are currently expected to last about a century, on average, before they are replaced. This means that there is a high probability that the original purchaser of any particular property will, at some stage, want to re-sell it. Second, real-estate properties are also far more expensive than almost all other commodities, so the eventual resale value is a very important matter.

Everyone who buys a property has to be concerned about its likely economic performance over time.

In the private sector, then, producers and buyers alike are powerfully affected by economic rules. If they break these, then sanctions of bankruptcy await them. But what about State agencies? State agencies, after all, produce a certain amount of urban form themselves. And through town planning processes, they also affect what private sector developers do. With the recent worldwide reduction in welfare state activities, the most important State involvement in producing urban form concerns the infrastructure, such as major roads, which are so important in the urban milieu. Infrastructure decisions, of course, are made partly on social grounds, but there is a powerful economic component here too. Indeed, with increasing global and regional economic competition, State decisions about the allocation of resources are increasingly driven by economic considerations of the most competitive kinds.

The same competitive constraints affect the practical operation of town planning controls. Here too, State agencies are often forced to accept that they need development more than the developers need them. Faced with the 'OK, we'll invest in South Korea instead' line, from increasingly footloose capital, they too are gripped in the pincers of profit-orientated economic rules. And they are often backed in this by voters who become more and more worried about such issues as unemployment and falling property values, particularly in the de-industrialising rust-belt areas which form so prominent a feature of those countries which came early on to the industrial scene. Here, for example, is what Detroit's Mayor Coleman Young has to say on the subject:

> Those are the rules and I'm going by the goddam rules. This suicidal outthrust competition among the states has got to stop but until it does, I mean to compete. It's too bad we have a system where dog eats dog and the devil takes the hindmost. But I'm tired of taking the hindmost.[1]

States, then, have commonly grown weaker in terms of their capacity to control local and national economies,[2] and in practice we typically find some kind of alliance – if often an uneasy one – between the profit-orientated private-sector developer, state agencies at the local and national level, and many voters too – that 'alliance of classes, structures and social forces' which, for shorthand, Stuart Hall calls 'the power bloc'.[3] The power bloc typically supports private-sector developers, and therefore helps them make a powerful impact on the vocabulary of types in good currency and on the wider production cultures through which these are defined as good design.

We must not, however, drift into economic determinism here: none of this means that either the typological vocabulary or the wider aspects of production culture are completely determined by the private-sector developer's economic requirements. The typification process is ultimately a process of negotiation and struggle, in which all the protagonists have at least some power, no matter how limited, to affect the outcome on the ground. As I know from my own experience, for example, well-organised community groups are often far from powerless, whilst design professionals always have at least some degree of autonomy in the form-production process, because they carry out complex work requiring personal judgement which it is difficult for others to control with any rigour. The power of those who might oppose the power-bloc, however, is limited; and the potential which it confers, in terms of working towards better-loved places, has to be carefully targeted if it is to make any significant impact. Attempts to use it in ways which run counter to developers' own profit-centred interests will be strongly resisted, so it is important to attempt this only where developers' interests really are inimical to getting better-loved places on the ground. If we are to avoid frittering away our limited resources in needless conflicts, we have to know as much as possible about which aspects of form-production culture support private-sector developers' interests, and so present likely barriers to change. That is what this second part of the book is about.

It is not too difficult to grasp the economic rules which private-sector developers have to follow, in capitalist situations, if they are to make enough profit to stay in business. There has been endless (often critical) discussion of these matters for more than a century, and the broad principles at least are widely known. But how do these economic rules relate to form-types and to the wider form-production culture within which these are situated? It is these far more difficult questions which are addressed in the chapters which follow.

As we saw in the last chapter, the typological level of urban change can be viewed as a process through which new types develop near the surface of production culture, generated in relation to a matrix of relatively stable, long-lasting 'deep' morphological elements, which are embedded in and guaranteed by legal institutions, and which constitute the basic 'building blocks' which make typological changes possible within the social system concerned. To understand the typological changes of the capitalist era, therefore, we must first grasp the nature of the morphological elements which underpin them. Amongst the long-running morphological elements which the capitalist form-production process took over from earlier times, the most deeply institutionalised are those which underpin and constrain the making of profits. Profit-orientated developers themselves are concerned primarily with those elements of built form which can be bought and sold: plots of land, and the buildings and associated outdoor spaces which are developed on them to increase their market value. If the production of urban form were left entirely to the efforts of particular profit-orientated developers, however, there is every likelihood that their individual impacts would lead to an overall situation whose unplanned nature would have unprofitable disadvantages for them all. From the overall profit-generation point of view, therefore, the attempts which central and local governments make to control developers' individual efforts, so as to maintain the competitiveness of the whole settlement in the global market place, are also crucial to keeping the capitalist development process going. Also of central importance, therefore, are the morphological elements around which government agencies focus their attempts to control development – the overall settlement which forms the area of jurisdiction itself, and the patterns of public space and land use within it. In this second part, therefore, we shall focus on the ways in which this range of long-running morphological elements is called on to generate particular types of settlements, public space networks, patterns of land use, plots and building developments during the capitalist era.

We shall start by exploring how the development of this typological repertoire is associated with the process of capital accumulation which lies at the heart of the capitalist economic system. Chapter 4, entitled 'Profit and place', will investigate how built form is involved in the accumulation of capital at two related levels: first as a commodity which is itself produced directly for profit, and second as a physical setting which affects the profitability of the production, distribution and exchange of other commodities of all sorts. The process of capital accumulation, however, must not be thought of as some abstract economic machine. It depends on human relationships, and is shot through with conflicts and contradictions which continually threaten both the future of the economic system as a whole, and within it the form-production process in particular. In Chapter 5, 'Propping up the system', we shall explore how the typological repertoire of urban form is influenced by power-bloc attempts to call on it in addressing these conflicts to keep the capital accumulation process going.

Overall, the picture we shall piece together will clearly show that new types have developed in ways which benefit the process of capital accumulation and help to reproduce its workings over time, but we must always remember that there is no inevitable, mechanical link between economic factors and the typification process: as we saw in the last chapter, it is particular human beings who fuel that process with the innovations from which new types might be abstracted. In the capitalist development process, as we have seen, it is largely design professionals who initiate these innovations, which cannot in practice give rise to new types unless large numbers of other professionals are also prepared to take them up and use them in their work. There is, however, a problem here for the professionals concerned: though professionals gain clear employment advantages from collaborating in the generation of new types to support the power bloc, many recent concrete results of this process on the ground do not seem widely loved by their users.

This creates a situation which is fraught with conflicts for the professionals involved, since it is an important tenet of professionalism that professionals are supposed to work at least partly for the broader public good. Those who continue to collaborate with the power bloc, therefore, have somehow to make sense of their work through particular ideological supports. These ideologies are crucial in maintaining whatever professional support there might be for the unloved status quo, so we shall have to overcome them if we are to make any serious attempt to get better-loved places on the ground. We cannot hope to overcome them unless we first understand how they work in making sense of professional life. In Chapter 6, 'Building bastions of sense', we shall therefore conclude this second part of the book by exploring the ideological material through which the design professionals who operate the nuts and bolts of the form-production process make sense of continuing to support it.

By the end of the next three chapters, then, we shall have a working knowledge of how recent typological transformations, together with their wider ideological supports, benefit today's private-sector developers. This understanding will enable us to see what we are up against if we want to challenge the status quo, helping us to target our limited power in the most effective ways.

4 Profit and place

If we want to understand how built form supports the current holders of economic power, in capitalist situations, we have to start by exploring how capital is put to work in the economic system. In as near to a politically 'neutral' definition as we are likely to find, the *Oxford English Dictionary* defines capital as 'wealth in any form used to help in producing more wealth'.[1] To understand how wealth itself can be employed to produce more wealth, we have to explore what is sometimes called a 'capital accumulation process'. In essence, this process can be thought of as a series of transformations. First wealth (in the form of money) is converted into supplies of raw materials and labour, bought as commodities in the market-place. Second, these are then converted into some other saleable commodity. Third, this is then converted back into money by taking it to the market for sale.

For this process to be profitable, the final amount received from sales has to be greater than that with which the process began. When this happens, so that the process concerned is profitable enough to be worth setting up as an on-going enterprise, the original linear chain of transformations becomes a recurring cycle. This is the 'capital accumulation cycle'; following the lead of Peter Ambrose,[2] we can represent it in a diagram (Figure 4.1). This cyclical process has an important time dimension. The more often capital is cycled through the system the more it can grow. The quicker each cycle can

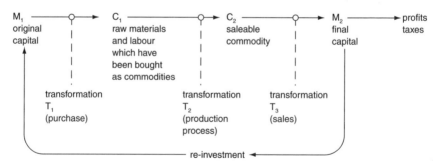

Figure 4.1 The capital accumulation cycle.

be completed, therefore, the faster the stock of capital will grow, giving the enterprise concerned an advantage over slower-cycling competitors.

Built form has direct effects on the speed and cost-effectiveness of the three key transformations in the capital accumulation process at two related levels. First, it forms the physical setting for the production and sale of all sorts of commodities, acting as what Henri Lefebvre calls 'productive apparatus of a giant scale'.[3] This gives built form a potential economic value in the capital accumulation process, which in turn creates the opportunity, at a second level, for producing built form itself as a commodity which can be traded in the marketplace, and for developing the profit-orientated businesses of construction and property development through which so many urban places are produced.

When they make market decisions, those with capital to employ in these built form businesses will be drawn towards the particular working practices and the particular aggregate settlement patterns, public space networks, plot developments and interfaces which they see as offering the best opportunities for capital accumulation. It is these, therefore, which become abstracted to the level of 'types' within development culture, and developers will defend them against less profitable proposals. To understand which types are likely to be abstracted and defended in this way, we have to explore the relationships between built form and the capital accumulation process in more depth.

The potential for capital accumulation in this process of property development arises from its three key transformations. The only potential for built form to affect profits, therefore, comes through any impact it may have on how these transformations are made. Of course, any given pattern of built form – or any working practice involved in producing it – might affect all the transformations in the capital accumulation process: the transformations themselves are not discrete 'things', but merely significant moments in a continuous process. Proposals for forms or working practices which have a disadvantage in terms of the cost effectiveness of any particular transformation (for example by costing more than their competitors, in materials or labour) may therefore still be implemented by developers, if they can eventually be sold for a disproportionately high price to offset these extra costs. It is the cost-effectiveness of the overall 'package' of transformations, rather than that of any particular transformation in isolation, which is the key to understanding why profit-orientated developers choose one particular form or working practice rather than another.

In this chapter we shall explore how physical forms, considered as commodities to be traded in the marketplace, are affected by the developer's search for the most competitive ways of making the three key transformations. As an analytical device for grappling with this highly complex process, we shall consider the three transformations one by one. First we shall explore the initial transformation from money into land, labour and building materials. Then we shall go on to investigate how these are themselves transformed into saleable building complexes. Finally, we shall analyse the way these complexes are turned back into money in the marketplace.

In the first of these transformations, the materials and workers needed for property development have to be brought to building sites which are geographically fixed. At the dawn of the capitalist era, pre-industrial transport limitations meant that labour and raw materials could only be purchased within a small geographical area near to the building site itself. Developers were therefore able to buy these prerequisites of production only in restricted markets, and in consequence there were only limited opportunities to shop around. All things being equal, any method of widening the area within which land, materials and labour could be purchased, for example through transport innovations, would be attractive to developers. This wider marketplace would increase competitive pressures on landowners, workers and the suppliers of raw materials alike, which in turn would lead to reduced unit costs for producers. Provided the cost of the transport innovation which created the opportunity for these savings was less than the savings themselves, then the innovation concerned would find a ready market.

Given the business opportunities this created in the transport field, it is hardly surprising that the last two centuries have been marked by an explosive proliferation of transport and communications innovations, from canals through railways, buses, automobiles and aircraft to the fax machine and the information super-highway. These increasing links between profitable production and processes of innovation in communications technology have had dramatic impacts on the repertoire of types through which urban form is generated.

The first of these impacts is felt through the effects of transport innovations on the availability of the building land which, as the fundamental raw material of the property development process, gives real estate its unique character as distinct from other commodities. In order to survive in the capitalist marketplace, most landowners have to maximise their productivity in competition with one another. With the expansion of communications possibilities, markets for saleable buildings are created at ever-increasing distances from established centres. Each new communication innovation creates the potential for a further spread of development wherever demand exists, or new demand can be created. Landowners and land buyers alike are therefore attracted to ideas for settlements with ever-greater levels of dispersion. The pattern of land use at a smaller scale, within the settlement, has also been transformed through this market dynamic. In the competitive marketplace, landowners are only prepared to sell land to the highest bidder. Any developer wishing to acquire a particular site must therefore bid higher than the competition, whilst still ensuring that the development will generate an adequate profit overall.

Specialised as they are, however, not all developers can afford to bid to the same level. The price which can be paid, whilst still achieving an acceptable profit, depends on the particular developer's estimate of how much the finished project will fetch in the marketplace. In turn, this will depend on a forecast of the amount any likely purchaser might be willing to pay in the location concerned. For any particular location, different developers – specialised in the production of space for different types of purchasers – will make different forecasts. There is likely to be a considerable disparity between what (say) a well-to-do house buyer and a small manufacturer seek in terms of location, and this will be reflected in the different amounts they will be willing (or, perhaps able) to pay to locate in any particular place.

The price which purchasers are able to pay depends on their economic situation: typically, for example, profitable companies or rich individuals will be able to outbid their poorer counterparts for a given amount of space. All things being equal, therefore, only the richest of all the users who might want to locate in a particular area will in practice be able to do so. Over time, this has led to a marked change in the pattern of urban land use.

In pre-capitalist situations, this pattern was highly constrained by the reliance on feet and on animals for transport. Everything had to be near to everything else, which meant that there was no practical alternative to a fine grain of mixed land uses: housing near work, shops and cultural facilities. In contrast, the dynamic of the capitalist land market, set free by the increasing mobility made possible for those who could afford transport innovations, led over time to an ever-broader zoning of land uses, so that purchasers of space who had different levels of buying power become more and more segregated from one another within the overall settlement fabric.

The practical logistics of the way land is acquired for development also has important implications at the smaller scale of the individual development site. Developers seek economies of scale in land acquisition; they are drawn towards the largest suitable sites they can afford. As their available capital increases, through the successful operation of the capital accumulation cycle over time, there is therefore a tendency for the average size of development sites to grow. This has important effects on all the morphological elements with which we are concerned. First, it brings about changes in the walls of public space. In pre-capitalist cities plot widths were typically narrow, so that as many plots as possible could take advantage of the accessibility offered by the network of public space; and each plot was occupied by a different building. There were thus frequent entrances on to public space, and also a high level of visual complexity as each building differed, however slightly, from its neighbours.

With larger plots and buildings, these characteristics change. First, since most buildings have only one entrance each, any given length of wall now has fewer entrances, which reduces the liveliness of the public space edge. Second, any given building now has a wider frontage on to public space. Since the distinction between one building and another is usually very noticeable, this too changes the character of the walls of public space, which become simpler in their visual organisation (Figure 4.2). More generally,

Figure 4.2 The visual impact of increasing plot sizes: Queen Street, Oxford; 1936
(top) and 1983 (bottom).

increases in plot size open the door for transformations in site layout. Small
sites, hedged in by their neighbours, permit innovation only at the scale of
the individual building; issues of how public access should be arranged, for
example, are already fixed. As sites get larger, there are still fixes round their
edges: any new development has to join up with the rest of the world, which
is beyond the developer's own control. Larger sites, however, have a greater
proportion of inner area which is under the developer's control, as compared
with edge which is not. This allows for innovations far beyond the scale of
the individual building; for example to encompass new ways of structuring
the public spaces which are required within large sites. Effectively, the
practical constraints which bound cities for millennia to particular patterns
of urban form become at least partially dissolved by the ability to seek for
profit at this larger scale.

One of the most powerful factors governing the way these new opportunities have been taken up in practice is concerned with increasing the

cost-effectiveness of labour. This is crucial for profitability, because labour costs are embedded in the costs of all the other commodities which are used for production: the costs of extracting and processing raw materials, and of making machinery and buildings to house production processes, in addition to the direct wages of those who carry out the processes themselves. If we add up all the cumulative labour costs involved, we shall find that they form an extremely large proportion of all the costs of production; some theoreticians of the left, indeed, would say that all these costs can ultimately be reduced to the costs of labour.[4]

Labour costs, then, are crucial for individual companies; but it is important to realise that in these days of global competition they are also crucial at the level of the State, because of the great influence of labour costs on the market competitiveness of the products produced in any nation state. In this situation it is vital, for the market competitiveness of any state, or any enterprise within it, that labour operates as cost-effectively as possible.

Because labour costs are so important, developers constantly seek new working practices to gain competitive advantages. When labour innovations are offered in the marketplace, therefore, those which are most cost-effective in developers' terms are likely to be replicated, eventually to become typical. This encourages a continuous economic rationalisation of the services on offer, through a process of ever-increasing specialisation, in which broad and complex working tasks are split down into ever-narrower parts, enabling a greater degree of specialised expertise to be applied to each particular aspect of the development process. Developers themselves are not immune from this pull towards specialisation. Increasingly, they produce space for particular specialised market sectors – particular types of housing, offices, shops or whatever. Within the complex division of labour which is generated through this process, the skills of innovation are particularly important to developers: faced with intense competition, they are always open to new ideas. This means that there are potential profits to be made from putting innovations on the market. This potential, in turn, sets the scene for working practices geared specifically towards innovation in the form-production process.

This is very different from the pattern of pre-capitalist working practices, in which the majority of run-of-the-mill proposals for built form were put forward by craft workers. The ability of craft workers to innovate was restricted by the fact that they made little distinction between what we would nowadays call 'designing' and 'making' in their work. Existing types of buildings, structural systems and constructional details were inextricably interlocked, and were embodied together in the worker's repertoire of practical skills. If major innovations were proposed, the whole package had to be relearned.

This did not mean that *no* innovation was possible. There was always the potential for innovation in small details, which therefore became the focus of aesthetic attention in this vernacular process. Innovation was needed at a larger scale too: since each particular building site must to some extent be different to all others, craft builders were constantly faced with new situations, and had to adapt their interlocking package of types from site to site. Overall, however, the potential for innovation in this process was limited by the fact that all the information used was tacitly embedded in the types of forms and working practices themselves – there were no abstract 'principles', except the minimum needed for site-to-site adjustments. The prerequisite for more rapid innovation, with its inherent market appeal to developers seeking competitive advantage, was the abstraction of these principles from the tacit working practices of the craft approach, and their ever-wider extension, to develop a new working practice in which 'design' became conceptually separated from 'making'. Those who developed such a skill found a ready market amongst developers, so the history of capitalist property development is marked by the ascendance of 'designers' to prominent positions in the form-production process as a whole.

The first designers, in this formalised sense, were architects; but they, of course, were not immune from pressures towards offering ever more specialised and rationalised working practices in the competitive marketplace. As each new service was offered, it had to be sold to potential buyers as something different from all the other services already on offer – it had to develop its own 'unique selling point', emphasising the factors distinguishing it from its competitors. Through this process, various architects found markets for more specialised expertise; giving rise to new types of engineering, surveying and so forth.[5] To maintain differences between these new specialisms was easy enough, because they covered different areas of expertise. It was more difficult, however, to distinguish them from the more 'generalist' approach of architecture, their original parent, since architecture had always included some degree of expertise in them all. The distinction here was made through emphasising the new practitioners as 'experts', as distinct from the generalist 'artist' architect – a move with which architects had to concur, to preserve a unique selling point for themselves in this competitive situation.

For their own economic survival, experts and artists alike have to offer their products and services in as wide a marketplace as possible. Calling on the growing sophistication of communications technologies, as important here as in other sectors of capital accumulation, and taking advantage of opportunities offered first through imperialism and later through the development of a global economy integrated beyond the borders of former empires, those designers who have proved most successful in the marketplace – whether as experts or as artists – have progressively shifted their radii of action from a local through a regional to an increasingly global scale. This shift has led to important changes in the nature both of expertise and of art, as designers in all fields seek knowledge which can be used to underpin design practice anywhere around the world, creating a demand for increasingly general and abstract design cultures, uncoupled from particular local traditions of art or expertise, and from such considerations as local climates and building materials.

The disembedding of this global design culture from local particularities makes innovation easier in every local situation. Though profit-orientated developers need innovations for their own market survival, however, they do not want *everyone* in the development process to be innovative, because of the inherent difficulty of controlling creative workers who have complex skills. To make the process easier to control, developers are therefore drawn to new working practices which call for less creativity from those workers who are 'makers' rather than 'designers'. This attraction is reinforced by the fact that less skilled workers are more easily replaced, and can therefore command only lower wages in the marketplace. The upshot of all this is that the creativity demanded from the designer is counterposed to a radical reduction in the creativity which is required from most other form-production workers.

Once under way, this deskilling process gains momentum by the way it interacts with the typification process. First, designers invent new forms which do not require craft skills. These forms are then bought in the marketplace. In this new context it becomes less and less rational, in economic terms, for workers to invest the time and effort required to acquire these skills in the first place. In turn, this makes it more difficult for any designers wanting to use craft skills in their buildings to find workers who can carry them out. By the 1920s, for example, the British architect Laurence Turner was bewailing the fact that,

> The average masons who frequent builders' yards in London are without experience and without knowledge of their craft, except in the performance of the simplest of their duties. To give them fine masonry to do would be to court disaster. It is not their fault, the work is not required from them. They lack experience.[6]

The effects here were compounded by the fact that designers, themselves competing with one another in the labour market, were under constant pressure to reduce their fees. This made it more and more difficult for those designers who still wanted to use a 'crafts' vocabulary to take over, as part of their own work, the detailed design decisions which could no longer be made by skilled craft workers.

More and more designers, therefore, became attracted by typological innovations which did not require such skills; so the demand for them fell still further, in an accelerating spiral of deskilling. The effects of this process at the typological level were geared-up by the design media of magazines, museums and so forth. Tom Wolfe recounts a story by the American architect Henry Hope Reed which illustrates this process very well:

Henry Hope Reed tells of riding across West Fifty-third Street in New York in the 1940s in a car with some employees of E.F. Caldwell and Co, a firm that specialised in bronze work and electrical fixtures. As the car passed the Museum of Modern Art building, the men began shaking their fists at it and shouting: 'That goddam place is destroying us! Those bastards are killing us!' In the palmy days of Beaux-Arts architecture, Caldwell had employed a thousand bronzeurs, marble workers, model makers, and designers. Now the company was sliding into insolvency, along with many similar firms. It was not that craftsmanship was dying. Rather, the International Style was finishing off the demand for it, particularly in commercial construction.[7]

This interaction between economic pressures on the designer in particular, and the deskilling of the workforce in general, has led over time to a radical shift in aesthetic focus. In the vernacular process, where innovative creativity could only be focused on the small-scale details, it was these which received much of the craft worker's aesthetic attention, leading to a rich variety of detailed expression within relatively constant building types. With the shift from craft to design, however, innovation could be focused on the whole building, whose overall form now received the aesthetic attention which had formerly been restricted to the details. With cost-restrictions on the production of large-scale drawings, it became ever more difficult to focus aesthetic attention on the small details, which therefore became considered effectively as by-products of the whole, relevant largely to the technical rather than the aesthetic sphere. This dynamic fostered (and was, in turn, reinforced by) an ever-increasing simplification in the vocabulary of detail-types (Figure 4.3); generating designs which require the production of the minimum number of drawings at the smallest feasible scale, within a rationalised, rectangular discipline – drawings which can

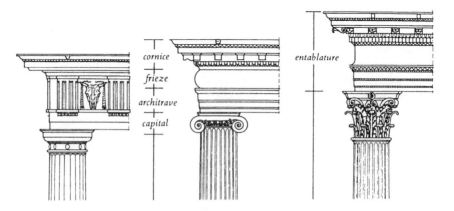

cornice

frieze

architrave

capital

entablature

Figure 4.3 The 'classical' typology of details: types now largely redundant.

readily be produced on the drawing board or through computer graphics systems, with the minimum of expensive hand work.[8]

As we can see from all this, the developer's search for design innovations to improve the cost-effectiveness of the first transformation from money into labour, land and building materials clearly has radical implications for all types of labour. These implications feed through to affect not only built form itself, but also the relationship between all those involved in the form-production process on the one hand, and on the other hand those who will eventually use the results of their work on the ground. Before the rise of the speculative development process, during the period when most buildings were produced for particular known purchasers, there was a very direct, personal relationship between each purchaser and all those involved in the form-production process. With the onset of speculative markets, this link became increasingly indirect. Except for rich people building houses for their own occupation, for example, the process of innovation was now controlled primarily through considerations of market competition, and the acceptability of any particular innovation came to be judged almost entirely in terms of its capacity to help the developer survive in the competitive marketplace. What we see here, then, is the gradual opening-up of a 'producer–consumer gap', with important consequences which we shall later explore in more depth.

Clearly, we can now see that the developer's drive to spend money on labour in the most profitable way has a wide range of important consequences for the form-production process. But what implications does it have for the other, inanimate commodities such as building materials which are also involved in this first transformation?

When they (or their designers) choose materials in the marketplace, developers are attracted to those which they regard as having the best balance between on-site cost and sales impact. This does not necessarily mean that they will buy the cheapest materials, but it does mean that they will want to get any given material as cheaply as possible. The purchase price of materials at their point of production is crucially affected by two related factors. One is the labour cost of producing them. The other is the economies of scale which come from large-scale production. The developer is therefore likely to be attracted by materials which are produced by cheap labour in large quantities. As we have seen, however, this attraction is tempered by transport costs. Building sites are fixed in their geographical locations. The sites where building materials are produced, from factories to quarries, are also fixed; so materials have almost always to be transported from where they are produced to where they are used. Since even the lightest and most compact of building materials are heavy and bulky by comparison with most other commodities, transport costs have important impacts on the cost-effectiveness of the materials concerned.

In the early years of capitalism, low technology transport led to delivery delays and to transport costs which were necessarily high in relation to the cost of materials at their place of production. These factors usually limited developers and designers to locally-produced materials, so the 'zoning' of materials took place at a fine grain, often with much variation between one relatively small region and its neighbours. As innovations in transport technology came on-stream during the capitalist era, bringing the potential for reductions in transport costs, developers were enabled to draw materials from ever wider areas; to take advantage both of the availability of cheap materials-production labour in particular areas, and of increased economies of scale in production. This has led to a situation, fostered by the increasing globalisation of design culture which we have already remarked, in which the materials produced in the largest quantities, with the greatest econo-mies of scale in their production, are likely to be used over large geogra-phical areas. At its extreme, this has fostered an internationalisation of materials use, on the one hand concentrating the production of high added-value components in the so-called First World, and on the other effectively buying cheap labour from the developing world for first-world building sites.

In rich countries, even the smallest of projects is nowadays affected by this global market. Whilst writing this chapter, I have been helping my son refurbish his house. In this tiny Oxford project, we have used rooflights from Denmark, plasterboard from Germany and Ireland, tiles from Thailand and Chile, and timber from Russia. In no case were these materials selected for exotic appeal – they were just the cheapest, reasonable quality items available. In larger projects, however, further economies of scale are possible if designers draw sparingly on this international palette of materials. Larger discounts can be achieved through buying large quantities from a small range of materials, rather than small quantities across a wide range; and this gives economic advantages to schemes which employ few materials.

Overall, then, the shift to a capitalist form-production process has carried with it radical changes in the way building materials are employed; with implications both at the largest and the smallest scales. At the scale of the overall settlement, we have seen a shift from locally-produced to globally-produced materials. At the scale of the individual building, however, materials are usually chosen from this wide range in restricted ways. This gives rise to a shift from a pre-capitalist situation in which, with rare exceptions, all buildings within a given settlement used the same materials palette, to one in which neighbouring buildings can differ widely in the palette they employ. On the one hand, large-scale variations disappear; on the other, variations between particular buildings intensify. Everywhere becomes more and more varied, just like everywhere else.

To summarise, this analysis of the transformation from money into land, labour and materials – the first transformation in the capital accumulation cycle – has yielded a range of useful insights into the dynamics of profitable form. We have seen how this transformation entails a radical division of labour within the production workforce, associated with a separation between 'designing' and 'making'; entailing a polarisation of creativity in work-patterns as designers – specialised as 'experts' or 'artists' – focus on innovation as a core skill, whilst a radical deskilling affects the tasks of others. We have noted how the materials used in construction have shifted from local to global in their geographical origins, but with only a restricted range being used in any particular building project. We have seen how settlement forms have become ever more dispersed, and how land use patterns have moved from a fine grain of mixed uses to a broader zoned pattern of segregation. And we have registered the ever-increasing scope for radical innovation which comes from the increasing size of development sites. All this is very revealing. Let us probe deeper, by analysing the second key transformation in the form-production process: the transformation from labour, land and materials to the finished building complex.

As long as capital is locked up in land, materials and labour, it is not circulating through the capital accumulation cycle, so it is not generating a profit. It is therefore extremely important to the developer that the process of converting these commodities into a saleable product should take as short a time as possible. This requirement for rapid construction has important form implications.

First, it supports innovations which allow as much construction as possible to be carried out under efficient factory conditions, rather than on messy and inconvenient building sites, often at the mercy (at least in the countries where capitalism began) of the vagaries of the weather. In principle, this draws developers towards innovations which employ pre-fabricated, factory-made components. The attractiveness of such components, in developer's terms, is increased by the fact that they hasten the process of craft deskilling. From the mid-nineteenth century, for example, the possessors of craft skills found themselves competing with machine carving and pressing techniques, even in a period when developers and their designers still wanted to use the detailed vocabulary associated with a craft-based construction process. Clearly, this accelerated the relative decline in craftsmen's wages, and gave another twist to the deskilling spiral.

Factories producing prefabricated components, however, are inevitably limited in number, relative to the number of building projects which might potentially use their products. This disparity is increased when component-factories, themselves capitalist businesses seeking their own economies of scale, increase in size but decrease in number over time. Because both factories and building sites have fixed locations, prefabricated elements will often have to be transported long distances to the sites where they are eventually used. This transport process is disproportionately expensive in the case of large, heavy elements; so in practice small components such as tiles, bricks, lintels, doors, windows, staircases, cladding systems and the like, are far more common than large assemblies. Given this restriction on the scale of prefabrication, it becomes doubly important, in profitability terms, that the on-site constructional processes should be based as far as possible on simple, rationalised assembly techniques. Again, innovations which make this possible will have a double attraction to the developer, since simple assembly techniques can be implemented to an adequate standard by deskilled, easily-disciplined workers.

The developer's desire to use workers in the most cost-effective way supports design innovations which involve repetitive building tasks, where identical construction procedures can be repeated many times in the course of a single building contract. This has a double impact on economic efficiency. First, the repetition of constructional tasks, which can never be completely deskilled, allows workers to speed up production and increase productivity, thereby reducing unit production costs, by spending more of their working time higher up the learning curve for the operations concerned. Second, it permits easier control of the whole production process. The storage and handling of materials, for example, is simplified as the variety of materials is reduced, whilst management is eased if design separates the work of the various trades so that no worker is delayed by waiting for another to finish. Attractions like these were already obvious to many actors in the development process by the beginning of the twentieth century: 'Economical considerations require that as far as possible there should be a repetition of parts', as Cornes put it, in relation to housing development, as early as 1905.[9]

If developers are attracted to innovations which generate these extra profit-potentials, architects have an equal motivation to propose them in the first place. This was already clear by the 1930s as the British architect H.S. Goodhart-Rendel saw:

> if we wished to build now in [an] informed and unhurried manner, we should find its cost prohibitive, not to the employer but to the architect. Just as in building itself, our methods have changed owing to the enormously increased cost of labour in relation to that of materials, so in . . . practice we now must save all we can of the principal's time and that of his draughtsmen if any profit at all is to be got out of the six per cent fee.[10]

Seventy years later, when the days of a six per cent architect's fee seem like a far-off golden age, these pressures are intensified indeed.

Taken together, all these innovations allow a given size of building to be constructed with less design time, and less time spent in the production of construction information, than would be needed for a more varied repertoire of materials and constructional tasks. Not surprisingly, the capitalist era has seen an ever-widening range of such innovations, abstracted into the repertoire of detail-types in good currency. Applied to ever-larger projects, as developers used the ever-growing availability of investment capital to gain economies of scale in production, this changed typological repertoire has further reinforced the reduction of close-up complexity endemic to the urban landscape in capitalist times.

To summarise, this analysis of the capital accumulation cycle's second transformation, from land, materials and labour into a saleable building complex, has given us further insights into the characteristics of modern urban form at an architectural level – an architecture of rationalised, deskilled construction techniques, using the minimum range of different details and materials in ever-larger projects. To complete this exploration of the form-production process, we have to consider the form implications of its final transformation – converting the completed project into money by selling it in the marketplace.

The marketed product consists of a plot or plots of land, with built fabric on it. Only these can be sold as commodities; and it is only these, therefore, from which profits can potentially be made. Public space, in contrast, cannot be traded in the market. Cities, however, are deeply involved with processes of communication and exchange; and even in these days of the information super-highway, public space is still a key medium through which these processes take place.[11] For the city to work in these terms, public space has to be freely available to all – free in the economic sense as much as (perhaps rather than) the political. If public space cannot directly be used to make profits, then it has no direct economic interest to developers. This means that it will be produced as a by-product of other commodities, such as buildings and cars, which *can* be used to generate profits. This 'by-product quality' was already noticeable to the Austrian architect Camillo Sitte, for example, as early as 1889:

> In modern urban planning the relationship between built and open space has been reversed. In the past, open space – streets and squares – created a closed and expressive design. Today the building plots are arranged as regular self-contained shapes, and *whatever is left* becomes street or square.[12]

Given the increasing potential for very large schemes, however, there *is* scope to make profits through attempts to commodify the spaces which connect together the various buildings within them. This process involves a shift from the highly connected grid of streets, typical of the pre-capitalist city, towards introverted 'enclave' spatial types such as culs-de-sac, malls, atria and the like, which can be given spectacular sales appeal (Figure 4.4), and whose common characteristic is summed up in the word 'exclusive' which is so often found in the real estate ads.

With all the building fronts facing inwards onto the enclave, this typological shift leaves only the backs of the development as a whole facing onto the outside world (Figure 4.5), a transformation which sets the final seal on the 'left over' quality of public space itself. Through the cumulative effects of this process, the public space network is transformed from a highly

Figure 4.4 Interior of the Westgate Shopping Mall, Oxford.

connected grid into a tree-like hierarchy, and the capitalist city as a whole is transformed into a series of islands, with spectacular interiors, set in a 'left over' sea. Within these islands, developers are interested primarily in building plots and what is built on them, and they have to sell these to make as much profit as they can. This means that they are drawn to innovations which seem likely to improve sales appeal in the speculative marketplace. Sales appeal must partly depend on prospective purchasers' individual preferences. Unlike bespoke producers, however, speculative developers cannot take these directly into account, since they cannot know the particular purchaser in advance. Speculative developers are therefore attracted to innovations which offer a wide and generalised market appeal, and are suspicious of idiosyncratic schemes. 'Yes, *you* like it, and *I* like it, but will it appeal to anyone else?' is the kind of question which will have been heard many times by anyone who has worked in this field.

Figure 4.5 The blank exterior of the Westgate Mall.

The desire to avoid anything idiosyncratic leads to a homing-in on 'market norms'; through the typification process, guided by market signals, which we already explored. To understand more about market norms, and to grasp the economic value they have for those who buy them, we have to investigate the common factors which influence buyers when they make their market choices. As has been pointed out, buildings have particular characteristics which make them different from other commodities which are bought and sold in the market. First, by comparison with other commodities, they mostly last much longer. Second, they are much larger: large enough to influence the workings of the capital accumulation cycle by forming the settings for the production, distribution and sale of most other commodities. Third, they mostly cost more. All these characteristics have important implications for the market choices which building-buyers make.

In today's dynamic social systems, the longevity of built form means that any building is probably far more permanent than its users' social arrangements. For instance, British home-owners moved house on average about every six or seven years during the 1980s, and even now do so on average every twelve, whilst any particular house may well last a hundred years. This means that the house might be lived in by anywhere between eight and sixteen different households, and the members of each, when buying it, know perfectly well that they will be faced with the need to re-sell in due course. The purchasers of buildings for other, non-residential uses – buildings which often cost even more – will usually have to face this situation too. Given the high costs involved, and aware of this need to re-sell, purchasers of all sorts of buildings will be attracted to forms which seem likely to hold their value, at least until the next resale takes place. Like the building's original developer, therefore, each of its successive potential purchasers during its long life will take into account the extent to which it conforms to their own concept of a widely saleable 'market norm'.

These characteristics of market norms are themselves strongly affected by their perceived suitability for housing particular uses. Where buildings are bought to house processes of production, distribution and exchange, purchasers will clearly be attracted by those which seem to offer the best balance between purchase price and promises of economic efficiency. The first of these promises is offered by the building's location, which affects the economic efficiency both of bringing raw materials and labour into production processes, and of sending finished commodities to the point of sale. This reinforces the importance of all sorts of communication and transport systems, and underlies the well-known dictum in development industry folklore that 'there are three key factors in profitable development: location, location and location.'

In deciding whether to purchase a building in a given location, entrepreneurs will clearly be influenced by its capacity to promote the economic efficiency of the production or sales processes which they want to carry on within it. Its influence on the cost-effectiveness of labour has particular importance here: as we have seen, labour forms a major proportion of overall business costs. Physical design has an important effect in this regard. Building layouts or arrangements of equipment which are inconvenient for particular processes of production or sales can slow workers down, whilst more convenient arrangements make it possible for the same tasks to be done faster. Developers are therefore drawn to innovations which increase cost-effectiveness: layouts moulded to support the particular processes concerned.

This has had a marked impact on the range of building types in good currency. Pre-capitalist settlements usually contained a relatively small number of highly adaptable types, in each of which a range of particular processes could, with some compromise, take place. This pattern has progressively been replaced with one which contains a far wider range of more specialised types, formed and named for particular uses. By the late nineteenth century, as the American architect Henry van Brunt noted in 1886: 'The architect, in the course of his career, is called upon to erect buildings for every conceivable purpose, most of them adapted to requirements which have never before arisen in history.'[13]

As the speed of change affecting processes of production circulation and exchange increases, however, there is a limit to the proliferation of specialised building types. The problem here is that the processes housed in any building now often change quite radically within the economic lifespan of the building itself, with the ever-present danger that the building's form might begin to reduce the efficiency of the processes within, rather than supporting them. Increasing dangers therefore arise when building interiors are tailored too tightly to any given pattern of activities. Adaptability and flexibility become watchwords now, and the vocabulary of building types in good currency represents the balance which is struck, at any particular time, between the specialised design which is needed to accommodate particular activities, and the flexibility required to allow for unpredictable change.

The third characteristic which distinguishes buildings from other commodities – their higher purchase price – has further implications for purchasers' choices. Many purchasers, requiring space but faced with this expense, prefer (or are forced) to hire rather than to buy. This is particularly true for the most expensive types of property, such as commercial buildings in central locations, where demand for scarce space in the best locations has pushed prices highest. Most purchasers in such locations, therefore, buy buildings as investments for letting to others. Developers producing such buildings are therefore attracted to innovations which have particular investment advantages.

In deciding whether or not to buy a particular building investment, potential purchasers will take account of two main factors. The first is the financial return: how much money will end up in the investor's bank account each year, after all the costs of getting it have been allowed for? But this is not all that matters: a second key factor is the financial risk involved. How widely will the building appeal to likely purchasers or tenants? Will the location keep its appeal? These are the sorts of questions which prospective investors have to address. And in answering them, to choose between buildings which generate the same level of income, investors will be drawn towards those with the lowest management costs and widest market appeal, in locations which seem most likely to improve their letting prospects over time. Since investors will pay more for such buildings, developers will be attracted towards them too. What do these characteristics imply for physical form?

When buildings are let to tenants, some management costs are always present. These are likely to rise, all things being equal, as the number of tenants increases. This is partly because there is more effort involved in finding the tenants in the first place, in drawing up leases and in collecting the rents. But it is also because of the greater likelihood of disputes over the day-to-day running of the building, particularly if the tenants concerned are engaged in different kinds of activities: 'how can you expect me to impress my clients when we get the smells from the restaurant kitchen all the time?'

Attracted by low management costs, potential investors are therefore drawn towards simple buildings with single tenants, rather than intricate ones with large numbers of small tenants involved in a wide range of different activities. If, through particular economic circumstances, no single tenant can be found to occupy all the space, it may in practice be necessary to let the building to a number of small occupiers. Even if this is known early in the development process, however, there is still an investment attraction in designing it so that it can eventually accommodate a single, larger tenant, should the opportunity arise in the future. The design, therefore, will be substantially the same as if it had been designed for letting as one unit in the first place. This creates further pressures towards the proliferation of single-use building types and towards market norms; but if all developers have the same market norms, why should any purchaser be attracted to one developer's product rather than another's? How can any developer steal a march on the competition?

The desire for positive appeal in the market, without falling into the trap of producing 'idiosyncratic' schemes, attracts developers towards forms which have their own 'unique selling point' – forms which stand out with a clear individual character, to be noticed by potential purchasers. To the extent that this can be achieved, to produce forms which are seen as 'unique' by potential purchasers, there is always the potential of selling them for a monopoly price.[14] There is a tension here, however. If the developer's product is too 'individual', it may put off prospective purchasers, who fear it may have too narrow a resale appeal. If it is too 'normal', then it runs the risk of not being noticed at all amongst its competitors. At one level, the profitability of development is about the balance between these poles.

In almost all urban cultures, from all historical periods, there are two common ways in which all kinds of 'important' buildings have been given forms which stand out from their neighbours. First, they are often physically separated from adjoining buildings, to take on a free-standing pavilion form. Second, they are often higher than the buildings around them. Developers seeking a unique selling proposition, with widely-appealing connotations of importance and prestige, are therefore drawn to the tall free-standing pavilion type when other economic factors permit (Figure 4.6).

Applied across a wide range of use-types, from family houses to office towers, this shift from connected building masses to the free-standing pavilion form has radical impacts on the character of public space. First, public spaces become less enclosed: not only are there gaps between the pavilions, but also they are often set back from the edge of the public space, separated from it by a band of private open space to set off their connotations of prestige. This makes it hard to perceive the public space itself as a positive 'figure' in the urban scene. Now it is the buildings which stand out as figures, against a neutral 'ground' of public space – that, of course, is the whole point of this shift, from the developer's perspective. At a typological level, the composition of the public space network changes from 'streets' and 'squares' towards a more generalised type of 'space'.

The story we have pieced together so far, through exploring the dynamics

Figure 4.6 A dramatic form of the city of pavilions: Hong Kong from the Peak.

of the capital accumulation cycle, has proved a complex one. To get into sufficient depth to make our exploration useful, we have had to make all sorts of analytical distinctions between factors which, in the real world, are linked rather than separate. In conclusion, we have to pull together the overall patterns we have found.

In the course of our exploration, we have uncovered transformations in the types of forms and working practices in relation to every physical scale within the overall settlement. The overall settlement pattern, for example, has become ever more dispersed, whilst the typical pattern of land use has been transformed from a fine grain of mixed uses towards larger 'zones' of single use, with increasing social segregation. The nature of public space has changed in two important ways: first from connected 'grids' to introverted 'enclaves', and second from 'streets' and 'squares' towards a more generalised type of 'space'. Building types have become more and more specialised in their intended uses, and the number of building types in good currency has consequently undergone a radical increase. In their physical massing, buildings have been transformed from constituent elements in a generalised, highly-connected mass into separate, free-standing pavilions, whilst the typical size of building project has radically increased. The interfaces between buildings and the public spaces which adjoin them has increasingly shifted from 'active' to 'passive' in nature. Finally, at the smallest physical scale, the surfaces of buildings have also undergone radical transformations. They have shifted from a 'local' to a 'global' use of materials, and they have changed from complex to simple, in terms of the patterning of the various elements from which they are composed. And, in parallel with all these physical transformations, making sense of them and reinforcing them, we have seen the rise of designers as 'experts' and as 'artists' operating with an increasingly generalised and abstracted design culture in an increasingly global marketplace, in parallel with the gradual deskilling of most other workers in the form-production process.

These typological shifts support the operation of the capital accumulation cycle, so the current holders of economic power will want to defend them against profit-reducing changes of direction. But we must not drift into economic determinism here; these typological changes are produced not by 'the economy' but by the very real people who are involved in the form-production process – a process which is shot through, in practice, with conflicts and contradictions. For anyone to make profits at all, the process which makes capital accumulation possible has to be reproduced through time, and this does not happen automatically. The current holders of power are constantly on the lookout for ways of helping this reproduction process to survive, in the face of the conflicts and contradictions which threaten it. Built form has the potential to help the system keep going, through its capacity to facilitate particular courses of action whilst constraining others. Powerful actors are attracted by types which have the capacity to help them here. For a fuller picture of capitalist urban form, therefore, we have to understand this dimension too.

5 Propping up the system

The capital accumulation cycle is a human system fraught with conflicts and contradictions, and it constantly threatens to grind to a halt from the profit-generation point of view. In order for profits to be made, it has somehow to be kept going through human effort. Those who feel that they benefit from the system, therefore, are continually attracted to ways of doing things which seem to have the potential for reproducing its particular set of rules and resources through time, by helping those in power to keep control of others who might prefer the system to change. It is not only in capitalist situations, however, that those with power are attracted to forms which help them to control others. Indeed, such attraction is a depressingly constant theme of urban history through the ages. What is unique to capitalism, however, is the fact that power-holders are drawn not to just any forms which might help them keep control, but only to those which also have the potential to help in the accumulation of capital, by addressing certain economic contradictions which are inherent in the capital accumulation process itself.

As we saw in the last chapter, the capital accumulation cycle embodies three key transformations: first from capital into land, materials and labour; second from these into a finished, saleable commodity; and finally back into capital through sale in the marketplace. In the first of these transformations, developers have direct control over decisions about the land, materials and labour which are bought – if things go wrong, they have only themselves to blame. The cost-effectiveness of the other transformations, however, relies heavily on the co-operation of other people. It depends on the willingness of consumers to buy the finished product, and on their ability to pay for it; and it relies on the efforts of the workers involved in the production process. In both cases there are obstacles to achieving this co-operation – obstacles which are embedded in the operation of the capital accumulation cycle itself.

There is, for example, an inherent contradiction in the cycle's third transformation, in which the finished product is sold in the marketplace. The contradiction here concerns the amount of money which consumers have available for buying commodities, to enable this transformation to take place. Clearly the amount of money paid to workers, in wages, has to be less than the amount for which the finished product is sold; if this were not so, there could be no profit at all. This means, of course, that the workers as a whole, considered as consumers, can never earn enough to buy everything they collectively produce.

From the producer's point of view, this generates a pervasive tendency towards under-consumption: the purchasing power available in the market-place is not enough to buy everything which is produced. This creates 'a pretty contradiction worth stressing', as the Belgian economist Ernest Mandel puts it:

> each capitalist always likes to see other capitalists increase the wages of their workers, because the wages of those workers are purchasing power for the goods of the capitalist in question. But he cannot allow the wages of his own workers to increase, for this would obviously reduce his own profit.[1]

In order to stay in business, those with capital are drawn to any approach which appears to have some potential to square this particular circle; the opening up of new, ever-larger markets – eventually on a global scale – through advertising, hire purchase, built-in obsolescence and manipulated cycles of fashion are obvious examples here. But approaches like these can only take effect to the extent that potential purchasers want to consume. If the capital accumulation cycle is not to break down altogether through under-consumption, therefore, there has to be a widespread and enthusiastic propensity towards commodity consumption. Where the cycle *does* keep going, this is exactly what we find: for many people, desire takes on a character which the sociologist Colin Campbell calls 'longing'.[2] Built form has the potential to reinforce and to channel the propensity to consume which longing promotes. Advertising, for example, can make a wider range of commodities seem *desirable*; but built form can be organised to make a wider range seem *necessary*, as we shall see. Urban forms which offer this prospect always have an attraction for those involved in accumulating capital.

Problems of under-consumption, however, do not represent the only contradictions within the capital accumulation cycle. There are also conflicts within the cycle's second transformation, in which labour and materials are converted into commodities for sale. So far as the economics of the capital accumulation cycle are concerned, the labour which has been bought for use in this second transformation is merely a commodity. But labour has characteristics which make it quite different from all other commodities involved in production. From mechanics to managing directors, the human beings whose labour is bought have thoughts, feelings and aspirations, organised according to rules which they have developed to enable them to make sense of their lives in capitalist situations. And there are tensions, in today's capitalist situations, between certain core rules and the very nature of work itself.

These tensions centre round the open, exploratory aspect of desire which, for many workers, is denied by the fragmented, deskilled nature of most of the production work which we explored in the last chapter. The tension here was nicely put by the French social theorist Raymond Aron, when he pointed out long ago that 'industrial civilisation subjects individuals to a strict discipline in their work . . . and yet it claims to have a philosophy of freedom, a philosophy of personality.'[3] Faced with such a contradiction, most work is likely to seem a boring chore rather than a fulfilling experience – something to be got through in order to earn the money to do more fulfilling things when *not* at work.

This might be good for promoting leisure consumption, but in production terms it is bad news indeed; it is essential for profit-generation that entrepreneurs get as much value as possible from the hours of labour which they buy. The clash between workers' values and the nature of capitalist work, which makes work seem a boring chore, is therefore extremely negative in this regard. For the sake of their own survival in the market-place, producers are drawn to types of working practices or built forms which seem to have a potential for addressing this problem. Most of these, of course, are concerned more with automation and with personnel management than with built form. But even the most sophisticated managerial approaches to 'job enrichment', for example, can do little to change the repetitive, deskilled nature of most work, without losing the economic advantages of the patterns which we have already discussed.[4] In this situation, management needs all the help it can find, to get the highest productivity possible from a bored and frustrated workforce.

Through the constraints it imposes and the opportunities which it opens up, built form has the potential to offer such help, and to offer it in an apparently apolitical way. Entrepreneurs are therefore always on the lookout for any built form innovations which appear to have the potential for increasing productivity.

So far we have argued that we might be able to understand many recent transformations in urban form, at least partly, as responses to endemic problems in the realms of consumption and production – problems which threaten the workings of the economic system as a whole. No matter how adroitly those in power choose forms to address these problems, however, the problems themselves remain, and they have widespread human impacts: with wages kept as low as possible – vital for market survival – many workers become frustrated consumers; whilst their work itself offers little but boring chores. In any situation like this, where the experiences both of work and of consumption are disappointing for many people, there is likely to be widespread discontent with the status quo. Those in powerful positions, who benefit from the very aspects of the status quo which cause the discontent, clearly do not want these aspects to change – they want to see the system

kept going more or less as it is, or changed in ways which benefit them still
further. There are consequently pervasive tensions between those who see
themselves as benefiting from the status quo and those who do not.

Such discontent is far from unique to capitalist situations: judging by the
recent demise of the so-called 'communist' regimes of eastern Europe,
discontents may indeed be far stronger in other systems. Widespread dis-
affection with the status quo does, however, raise particular and special
kinds of problems in societies which are organised along capitalist lines,
because of considerable limitations on the deployment of state power to deal
with any practical implications which might arise from it. Batons and
jackboots are brought out of the closet from time to time, but they do
not fit well with a 'philosophy of freedom'; so the widespread public consent
necessary to underpin such an approach is only grudgingly given. These
ideological limitations on the deployment of state power give built form a
particular importance as an 'apolitical' medium for addressing social discontent in capitalist situations. Both developers and State planning agencies
are therefore drawn towards built form innovations which appear to have
the potential to render discontent impotent in its practical outcome, and to
support top-down social control.

To summarise the argument so far, we have seen that there are two
important conflicts built into the capital accumulation cycle. On the one
hand, consumers can never be paid enough to consume all that they
produce. On the other hand, there are tensions between workers' aspirations
and the nature of capitalist work. As well as their economic impacts on the
process of capital accumulation, these two conflicts together give rise to
more general tensions in the political sphere. Because of its capacity for
supporting some actions and constraining others, built form has the potential to help keep the lid on these tensions – a potential which is conveni-
ently 'apolitical' in nature. Both developers and State agencies are therefore
drawn towards innovations which they see as positive in this regard. If they
prove successful in these terms, the innovations concerned are supported,
and eventually the range of types in good currency is transformed under
their influence. But how do these transformations develop in practice? That
is the question we must try to answer next. We shall begin by exploring
how built form can help in promoting consumption – since everyone
consumes, this is where we might expect to find the most pervasive
influences on the character of urban form.

Broadly, there are two ways in which built form can promote increasing
consumption. First, its physical layout can be arranged so that it is difficult
or inconvenient to use, according to commonly-held rules of daily life,
without the support of a great number of other commodities, which therefore come to seem necessary. Second, it can be laid out in such a way that
consumption can more easily take place individually, or in very small
groups such as the nuclear family, rather than at a larger collective level,
so that all consumers have to have their own widgets rather than a relatively

small number of widgets being shared. The history of capitalist urban form has been marked by the ever more widespread adoption of spatial types which have these effects, at the same time as promoting enhanced opportunities for those with power to control those without it.

In pre-capitalist times, overall settlement patterns of form and use were organised for low levels of consumption; with limited resources, this was a matter of necessity. Settlements were usually compact in their form; with movement of people and goods depending on the low-energy technologies of feet (human and animal) they had to be. Similar constraints on movement governed the settlement's land-use pattern. The spatial distribution of activities formed a fine-grained pattern of mixed uses, so that most people lived within walking distance of their working, shopping and leisure venues. With frequent reliance on shared water supplies and sanitary facilities, energy requirements for servicing were also kept at a minimum. All in all, both the form and the use-pattern of pre-capitalist settlements provided just about as unpromising a *parti* for promoting consumption as anyone could imagine. Developed through practical processes of trial and error, through centuries of low energy-availability, things could hardly have been otherwise. As capitalist processes developed, and as consequent problems of under-consumption became more and more evident – slowly at first, but gaining pace into fully-fledged international crisis during the 1920s – ideas for settlements whose form and content offered the potential for increased levels of consumption and social control, became more and more attractive to producers and to the local and national agencies of many states. Which types came to the forefront through this process?

Levels of commodity consumption are increased, all things being equal, as settlements become more dispersed, with a given number of inhabitants being spread over a larger geographical area. First, the more people are spread apart, the greater the consumption of the building materials which are required to construct the roads, drains and other linking infrastructure. Second, life in the new dispersed places requires the purchase of new commodities; as the physical distances between inhabitants become larger, taking people beyond sight, earshot and walking distance of each other, 'free' modes of communication, using the unaided body, become impossible. If settlements become more dispersed, therefore, people have to buy all sorts of new commodities to overcome the spatial separation involved. So far as the physical movement of people and goods is concerned, we can see that the transformation of settlement types towards greater degrees of dispersion has promoted the demand for all sorts of transport commodities. First, these were centred around public transport systems, with all their associated hardware, and the secondary commodities such as coal, oil, petrol, spark-plugs, tyres and so forth which they required. Later, the dispersed low-density settlement also helped to promote the shift from these collectively-consumed transport commodities towards the higher consumption levels associated with individualised transport forms, in particular the private car.

Once under way, this process has an in-built tendency towards ever-greater levels of dispersion, because cars require more space per person than, say, pedestrians or cycles – on average, about seven times as much as walkers, and at least twice as much as cyclists. As increased dispersal leads to more car use, therefore, a multiplier effect comes into play. To the greater level of car use has to be added still further traffic for the delivery of petrol and oil, and of items such as tyres and spare parts, to say nothing of the energy used in producing all these things. In turn, all this requires more space for such ancillaries as roads, petrol stations, garages, residential parking and car parks for shops, offices and factories. The land this uses up creates still more dispersal of other facilities, which generates in its turn a need for yet more car use, and so on and on in an ever-expanding spiral of dispersal, until an extremely dispersed settlement such as Los Angeles, for example, has some 70 per cent of its land area put to car-related uses.

The need for communication in the dispersed settlement, however, concerns more than just physical movement. It is also to do with the interchange of news and ideas between the people who live there; so dispersal also promoted a demand for innovative commodities such as the telegraph, the telephone and, later, all the bric-à-brac needed to travel through cyberspace on the electronic highway. Many other commodities also become necessities for the changed pattern of life which has now been set up, in the context of a host of other changes to urban form which dispersal brings in its train. One important change concerns the pattern of land uses. For example, dispersed development restricts the number of people who occupy a given area of land. This means that the collective facilities such as shops and leisure venues, which rely for their profits on selling to a certain minimum population, also have to be spaced far apart for their own economic survival. This makes it inconvenient and time-consuming for people to make frequent visits to the shops, for example; so there are pressures to buy everyday requirements, whenever possible, less frequently and in larger quantities at any one time. This in turn generates a potential demand for yet other commodities, such as storage cupboards, refrigerators, canned foods and tin-openers. And, of course, it also generates a demand both for equipment and for services to help dispose of all the new kinds of rubbish which this extra consumption creates.

Given the pressures to increase levels of consumption if the overall system is to survive, it is not surprising that transformations towards dispersed settlement patterns have proceeded apace in capitalist situations. Local and national State agencies, involved in processes of regional and global competition, have offered them legal and fiscal support, creating, for example, legal frameworks for credit and tax incentives for purchasers, and granting planning permission for projects based on the new types. Transformations towards dispersed settlement patterns are also attractive, in terms of keeping the system going, because of their ideological effects. As is well known, early capitalism was associated with high levels of urban social deprivation.

Thinkers of the time, both of the left and of the right, saw the importance to those in power of keeping this hidden, as far as possible, from better-off people, in order to avoid the rise of movements for social change which might reduce profits in all sorts of ways.

The new technologies of transport which proliferated as part of capitalism's early development were highly relevant to this ideological dimension of settlement form. On the one hand the large-scale transport structures, such as main arterial roads with their associated commercial facilities, could be used very effectively to conceal working-class misery. Friedrich Engels, the German political writer who carried out much of his work in England, pointed this out very clearly. Referring to what he called the 'hypocritical plan' of Manchester in the 1840s, Engels observed that well-to-do people,

> can take the shortest route [from their homes] through the middle of all the labouring districts to their places of business, without ever seeing that they are in the midst of the growing misery that lurks to the right and the left. For the thoroughfares leading . . . in all directions out of the city are lined, on both sides, with an almost unbroken series of shops . . . [which] conceal . . . the misery and grime which form the complement of their wealth.[5]

Thinkers of a less leftist persuasion, however, were more likely to see the concealment of misery and grime in positive terms, and to reserve their criticism for the occasions when the new transport technologies had the opposite effect of bringing it all too clearly into view. Here, for example, is another German – the influential town planner Joseph Stübben – writing about Berlin from the political standpoint of the Prussian technocrat:

> In most places, in older parts of the city, the urban railroad cannot assume a place in the middle of the street, but must instead cut right across the perimeter block system . . . Unfortunately, this type of construction has the disadvantage of causing great unpleasantness . . . for the passenger on the train, who, on his journey through the city, already has tasted a sequence of disgusting images composed of backyards and rear views of buildings, and has thus gained obnoxious insights into the miseries of metropolitan life.[6]

Poor passengers! As the artist Gustav Doré showed, with his telling engravings sketched through the other end of the telescope (Figure 5.1), this sort of spatial relationship was also none too hot for those who had to live in the backyards which had such 'disgusting images'.

From the point of view of maintaining the status quo, 'obnoxious insights' were better avoided, by using urban form to separate the classes where possible, rather than bringing them into contact. Viewed at the level of the settlement as a whole, the new transport technologies – which made

Figure 5.1 Gustav Doré: viaduct in London, 1852.

dispersal possible in the first place – also offered opportunities for separation, through the physical infrastructure which they required. In order to
operate without continual delays, the railways (and later urban motorways)
were often raised up on viaducts to avoid interference with ground-level
streets. These viaduct structures were perhaps no larger in scale than the
fortified walls which had been a feature of many earlier European cities, for
example. But now they ran *through* and *across* the settlement, instead of
around it, in the process dividing it perceptually into a range of separate
fragments, with a great deal of potential for making the poor invisible,
literally 'on the wrong side of the tracks'.

Given their double economic and ideological attractions to those with
power, transformations towards dispersed and fragmented settlement forms
have, in practice, received strong support both from private sector producers
and from State organisations in many parts of the world. At a smaller scale,
similar support was given to parallel transformations in public space networks. In the pre-capitalist city, limited transport opportunities placed
strong constraints on the layout of public space. Typically, in the areas

where capitalism began, we find that cities – necessarily organised primarily for pedestrian movement – had close networks of streets linking everywhere to everywhere else, with relatively straight links bringing people from outside any settlement right into its heart without the need for energy-wasting detours and backtracking, forming a pattern which has been termed the 'deformed grid'.[7]

Limited pedestrian travel distances had the further effect of encouraging intense competition for the most accessible land, so that the most accessible parts of cities of any size became densely developed. In turn, this had further implications in terms of public space; this was restricted to the minimum area required for its practical use, to the extent that public space which proved itself underused was taken over for other purposes. In Britain, for example, as open markets became less important after medieval times, many were taken over for the development of commercial buildings in the process of 'market colonisation'.

This discipline ensured that the remaining public space was intensely used. Given the very small amounts of private space and facilities which were available to them, the members of the 'labouring classes' continued the intensive use of public space into the early capitalist years, whilst the complexity and intensity of street life in the United States, for example, was still further increased by the multi-cultural origins of the immigrant population. The American sociologist Richard Sennett, for example, describes the teeming street life of Halstead Street, in the heart of the immigrant ghetto of Chicago, around 1910. In part, this was based on straightforward – albeit informal – commercial activity: 'it is amazing to see in old photographs of Halstead Street the young and old, shoulder to shoulder . . . shouting out the prices and virtues of their wares.' In part, however, this activity subverted conventional notions of law and order, in ways which the well-to-do would have found alarming indeed:

> Many youths would, with the tacit consent of their parents, enter into the more profitable after-school activity of stealing – we read, for instance, in the letters of one Polish family of great religious piety, of the honor accorded to a little son who had stolen a large slab of beef from a butcher on the corner. Life was very hard and everyone had to fight for their needs with whatever weapons were at hand.[8]

Public space arrangements like these had little to offer in terms either of consumption or of drawing attention away from social injustice. The deformed grid was a sight too convenient for the 'free' energy of the unaided human body, whilst the intense use of public space encouraged just the sort of face-to-face contact likely to make social injustice all too obvious, to the dismay and fear of those in privileged positions. Both producers and State agencies would be attracted by public space innovations with the potential to avoid these shortcomings – innovations which held out the promise of increasing consumption and promoting perceptions of social harmony. From the point of view of increasing consumption, the most obvious way in which the layout of public space can help is through its impact on the demand for commodified transport. In this regard, consumption can be increased by new public space types which promote the use of mechanical transport; particularly when this is individual rather than collectively 'public' in its nature. Both producers and State agencies, therefore, are attracted to innovations of this kind.

The effects of this attraction, particularly since the rapid development of the mass-produced private car, since the First World War, have changed both the form and the use of public space. The form has progressively been developed to promote the convenience of the private car, strongly reinforcing the typological transformation of the public space network from the grid to the hierarchy, which we explored in the last chapter. The main connecting spaces, high up the hierarchy, have on the whole been developed as wider streets with larger radii at corners and intersections. This increases the attractions of car-driving, making it easier at a given speed. In turn, this is good for the sales of cars and all the other commodities, from petrol to insurance, which are associated with them. But these transformations have a double attraction in terms of propping up the system overall. Not only do they help to increase consumption, but they also have advantages in terms of concealing social injustice: because the inundation of the main public connectors with fast-moving cars poses dangers to other potential public space users, streets with high levels of car use act as barriers to movement across them, dividing the city into discrete fragments. The restructuring of public space to encourage car ownership therefore reinforces the shift towards 'enclave' spatial structures, furthering settlement fragmentation and with it the increased opportunities for concealing social injustices which we have already discussed.

The effects of introversion here are still further emphasised by emergent impacts on the character of the major traffic streets which remain. If these main traffic distributors are to flow freely at (say) 30 mph – in practice, often at higher speeds – then access off them can only be allowed at widely-spaced intervals. It is certainly not possible to have a direct vehicular access to each dwelling, for example, without limiting the through-flow to a considerable extent. Since it is nowadays widely regarded as inconvenient to have a dwelling without vehicular access, and since this is usually regarded as the principal access, dwellings do not front on to the main traffic distribution roads, but rather turn their backs to these, to front inwards on to the slow-speed culs-de-sac lower down the traffic hierarchy. This means that the major distributors are usually faced with blank walls or fences (Figure 5.2), rather than having the active fronts of earlier times. So far as major connectors are concerned, this results in a radical transformation of the interface between plot developments and public space, from active to passive.

Figure 5.2 Blank walls to a main distributor road.

The extent of blank walls facing on to public space is still further increased when, as is often the case, the dangers of a car-focused public space network support the spatial segregation of pedestrians from car-drivers. Given high levels of car-reliance, the main entrances to buildings in segregated layouts are almost always on the car side, in practice if not by the designer's intention. Pedestrians, therefore, have to go 'round the back'. In the residential areas which make up most of the zoned-out settlement pattern, and where this segregated arrangement is usually found, privacy is important round the back in most cultures. In social terms, this is what the conceptual distinction between 'front' and 'back' is about. Public access paths at the back, therefore, tend to destroy the privacy of the building and its associated outdoor spaces. To overcome this problem, privacy barriers such as fences or hedges are erected – if not as part of the original layout, then later as privacy problems begin to make themselves felt. These barriers make a completely blank interface between pedestrian space and the adjoining properties, cutting pedestrians off from all signs of life, except those from other pedestrians in the same space (Figure 5.3).

This isolating effect is reinforced by the complex plan-forms which pedestrian spaces often take. Given the ability of pedestrians (unlike cars) to negotiate sharp corners, the (unsaleable) pedestrian spaces are often fitted into whatever locations are available round the back, so as to impose the

minimum constraints on the profitable layout of the saleable plots and
building. Often this has generated pedestrian spaces with a somewhat
maze-like quality, with no long views. Even if other people are present in
the pedestrian space – and many are now elsewhere, in cars – they are often
round the next corner, out of sight. If one were to set out purposely to
invent a public space pattern to isolate individuals one from another, this
would be it. There is little support for the development of social solidarity
here.

As at the larger scale of the overall settlement, then, the networks of
public spaces within it have also been progressively transformed in such a
way that they promote consumption at the same time as they make social
control easier by making bottom-up social solidarity more difficult to
develop. The logic of our argument suggests that we should find similar
transformations in regard to plots and buildings at a smaller scale. Is this
borne out in practice?

Figure 5.3 This is what happens when public space adjoins private outdoor space.

In pre-capitalist settlements of any size, the pressure towards high densities – at least near the centres of settlements – had important implications for what plots and buildings were like. Since technological limitations restricted the height of buildings, high densities could only be achieved through the development of continuous building masses, with each building in close contact both with its neighbours and with public space. This, in turn, reduced energy consumption, since it meant that each building helped to insulate its neighbours against adverse temperature changes. Together, the individual buildings reduced each other's heat losses and gains, so that the ensemble required the minimum external energy input to achieve the relatively undemanding comfort levels typical of pre-capitalist times. Internally, too, building layouts were well-adapted for low levels of consumption. Except for those of the most well-off, the interiors of buildings had few separate rooms. This meant that levels of internal privacy were low, which encouraged the various occupants to pursue a communal style of life, with relatively few commodities shared amongst them.

As at the larger scale, therefore, the structure of pre-capitalist built form at the smaller scale of the individual building was not a promising starting-point for encouraging high levels of consumption. For those worried about under-consumption, therefore, changes were desirable. Developers were attracted to innovations which called on opportunities, created by dispersal, to create high-consumption building types. For example, a shift from linked-up building masses to free-standing pavilion forms – made possible through the lower densities of dispersed settlement forms – carries with it many such opportunities. A large number of free-standing buildings creates, in aggregate, a very large area of roof and wall surfaces, gaining or losing heat to the external climate, so that the people who use these buildings have to consume disproportionately large quantities of commodities such as coal, coke, oil, gas, electricity and so forth to achieve a given level of climatic comfort.

As well as these transformations, which had a general effect on a wide range of building types, the search for forms to increase both consumption and the opportunities for top-down social control had specific effects on particular types. It is important that we understand these effects, since significant aspects of overall changes in urban form are brought about through the aggregate effects of changes at the building scale: in practice, capital is channelled into urban form largely through the medium of discrete building contracts. We shall explore the transformations which have affected three types of buildings. First, we shall consider buildings for work, in which the production of commodities, from buckets to advertising campaigns, takes places. Second, we shall analyse the transformation of shops, where most everyday commodities are purchased by their users. Third, we shall explore changes in dwellings, within which so much every-day commodity-consumption goes on. We shall find that all these building types have been transformed in important ways through attempts to address the endemic conflicts of the capital accumulation cycle.

In the context of buildings to house working processes, purchasers – and therefore developers – are drawn towards transformations which offer the prospect of increasing the cost-effectiveness of labour in the production process. As we saw in the last chapter, the physical layout of buildings has an obvious effect on the potential productivity of labour. Building layouts or arrangements of equipment which are inconvenient for the particular work process can slow up work, whilst more convenient arrangements enable it to be done faster. But physical design can only create the *potential* for faster work; it cannot by itself guarantee that this potential will be realised in practice, for the workers might prefer to use the time which has been saved for their own purposes, such as socialising with their workmates. Indeed, given the mismatch between the 'philosophy of freedom' and the nature of capitalist work, this is really rather likely. It is the sort of thing I constantly do myself, despite the fact that I have a job which allows infinitely more scope for choice and personal initiative than most people's do.

It is the task of management to ensure that the time saved through convenient layouts is used 'productively' rather than for the worker's own pleasure; and developers are drawn towards innovations in building design which seem to have the potential to aid managers' efforts. Management techniques come and go, but a great number of these approaches depend to some degree on the visual monitoring of what workers are up to. The importance of this surveillance was clearly understood from the beginning of capitalist production. It was given theoretical expression by the British philosopher Jeremy Bentham, for example, when he put forward his 'Central Inspection Principle' or 'Panopticon' – a way of organising the spatial structure of buildings in a radial formation, so that a single central observer could oversee the operations of a multitude of other people. Bentham's brother Samuel adapted this Panopticon principle to the design of an 'industry house'; but in practice its radial geometry did not fit very well with the sorts of spaces needed for the convenient arrangement of machines, belt drives from steam engines and so forth. A less elegant but more practical version of the same principle was achieved, from the earliest of the large capitalist mills, merely by having large open factory floors where all the workers could be kept in view by a small number of overseers.

The design and layout of equipment within the open production space is also clearly crucial to labour's cost-effectiveness, so entrepreneurs are constantly drawn towards innovations in equipment which can improve productivity. There is a double attraction in equipment innovations which can, at the same time, sharpen the potential for surveillance. The invention of the moving production line at Ford's Dearborn plant in 1917, for example, gave a further technique for ensuring that individual workers fitted into an overall production discipline in time as well as in space: those who did not were instantly exposed. Nearer to our own time, the introduction of computer controlled machinery gives an electronic version of the same ability to monitor each worker's contribution and, in this case, to make automatic records of individuals' patterns of work. As large corporate businesses were developed, requiring large numbers of office workers to co-ordinate operations, office space also took on the 'open floor' arrangement; first in the USA, then in Britain, and finally more widely after the Second World War. As early as the late nineteenth century, in the US, highly-surveilled offices – often presented with a democratic 'all in it together' rationale – were thought appropriate even for the most 'creative' businesses such as architecture; the offices of star Chicago firms such as Burnham and Root or Adler and Sullivan had already taken this form by 1890.

Shopping buildings too underwent radical transformations, as their purchasers and developers sought ways of increasing consumption which also supported top-down approaches to social control. Attempts to increase consumption here have centred around the creation of an ambience which first attracts more people to enter, and then encourages them to buy. The earliest of these attempts relied mainly on better arrangements for bringing the goods on sale to the attention of potential purchasers, most obviously through the use of the ever-larger windows made possible by the availability of larger sheets of glass from about the 1830s onwards. Later, new spatial types were developed for the promotion of increased consumption. The large internal spaces of arcades, department stores and shopping malls, for example, were given spectacular architectural quality. Because we are used to indoor spaces being far smaller than outdoor ones, the spaces of malls and arcades which might have seemed quite modest in scale had they been out of doors, took on the impressive character of palaces or cathedrals once they were roofed over.

As alternatives to the earlier shopping street, these spectacular spaces had many advantages in attracting consumers. First, unlike the street, it was always summer under the glass roof. Second, the character of what went on in there could be – and was – managed in such a way as to delight desirable shoppers, rather than relying on mere convenience for sales. As early as 1883, for example, the novelist Emile Zola likened a Parisian department store to 'a temple to Woman, making a legion of shop assistants burn incense before her'.[9] Nowadays, of course, the desired clientele would be wider than Woman, and the technology available for ambience-creation is different – muzak rather than incense – but the principle remains the same: management to create a spectacular ambience where the 'feel-good factor' encourages the optimism needed to spend to the hilt, borrowing to do so, even when bad times hold sway in the street outside.

Any optimism-creating ambience, of course, has to exclude signs of deprivation or conflict which might lead consumers to question the way things are going. The fact that the typical mall, for example, is privately owned makes it possible for its managers to exclude people and activities which might have a negative effect on consumer confidence. A 1994 survey of British shopping centres found that most of those reviewed were run in this 'exclusive' way, defining as 'undesirable' a wide range of activities such as overt political action, begging and displays of homosexual affection, which might be perceived as challenging the inevitability of the status quo.[10] This exclusiveness is made still easier to maintain when the mall is 'out of town' – a further extension of the dispersal principle. Now many potential 'undesirables' find it difficult to get there in the first place – the out-of-town location itself subjects visitors to a first level of economic screening, since the ability to gain access depends on the visitor being at least sufficiently solvent to have the use of a car.

Perhaps it is not surprising, given the widespread increase in all sorts of social tensions in many cities in many countries since the 1980s, that this exclusive strategy seems to be well-liked by many of the most free-spending shoppers. This raises a problem for high-street shopping, which is forced to compete with malls in the marketplace. This in turn has led to the increasing application of mall-type management strategies to traditional outdoor streets, through the new discipline of 'Town Centre Management', paralleled by local government taking on new legal powers through local by-laws. In my home town of Oxford, for example, there are proposals to make it illegal to drink alcohol in the street. Now, I am no more in favour of public drunkenness than anyone else, but it is surely obvious that sweeping streets 'clean' of all signs of social stress must reduce everyone's tendency to question the status quo, because it renders invisible the social problems which the status quo entails.

In ways which are perhaps less obvious, the search for methods of increasing the potentials for consumption and for top-down social control have also come to affect the forms of residential buildings, now developed, as we have seen, in dispersed settlement forms. Because of their low densities, these dispersed forms cannot contain high levels of social or entertainment facilities near to where people live. There is therefore an increase in the potential demand for commodities which enable commercial entertainment to take place in the home rather than collectively: the radio, TV and video rather than the cinema, the hi-fi rather than the music hall or concert venue. This shift away from the communal consumption of leisure facilities also makes each dwelling more and more self-contained, giving yet another turn to the dispersal screw.

At a range of physical scales, then, the development of capitalism has been supported by radical changes in the typologies of urban form. Particular forms have been selected in the marketplace from amongst the innovations on offer and have eventually attained the status of types, because of their double potential for assisting in capital accumulation on the one hand, whilst helping to increase consumption, manage labour and defuse public awareness of social injustice on the other. This market selection process has given a further push to the transformation of overall settlement patterns from compact to dispersed forms, whilst patterns of land use have shifted from a fine grain of mixed uses towards an assemblage of large zones of simple use. Public space networks have shifted from highly-connected deformed grids towards increasingly tree-like hierarchies, whilst the uses taking place within these networks have become more and more limited to those associated directly with transport, in a radical shift from the multiple uses of earlier times. Building types too have undergone radical change; with further pressures towards a general move from agglomerated masses to individual pavilion forms, whilst their interfaces with public space have been transformed from an active to a passive character. And all these transformations are now taking place across the world-wide range of an increasingly integrated global economy.

These typical transformations, so important to keeping the system going, do not happen by osmosis. They happen by design, and they therefore require the active participation of design professionals, who somehow have to make sense of supporting these typological shifts in the course of their working lives. This requires the development of particular kinds of subjectivities. In the next chapter, we shall explore the ideological material from which these are constructed.

6   Building bastions of sense

In the last two chapters, we explored the transformations which the typological repertoire of urban form has undergone during the capitalist era, in the service of both immediate and long-term capital accumulation. There is plenty of evidence to show that many of the results on the ground are widely unloved.[1] We know, however, that these transformations have not happened through some malign mechanical linkage between economic factors and the typification process: these new types could never have moved into the mainstream of production culture without the active collaboration of large numbers of the people involved in the development process. None but the paranoid, however, would believe that the people involved in the development process, helping to make these transformations, produce their negative impacts on purpose or would not avoid them if they saw that they could. Even speculative developers, demonised in the minds of many, would prefer to make better-loved places, if only to broaden the market appeal of their products. This must mean that many of those involved in the form-production process, unlike many of those who only use its results, must currently be able both to think and to feel that these transformations have largely been acceptable ones.

From the users' perspective, the cultural patterns which are embodied in thoughts and feelings like these, some of which we have encountered in passing in Part I, are counter-productive: they block the potential for those involved to use whatever limited autonomy they might have to work towards places which might be better-loved in users' terms. From this point of view, it is important to break free from these established cultural patterns. Before this can happen, however, we have to understand more about how and why they arose in the first place.

As a starting-point, we can see that the unloved nature of recent transformations places all the members of the development team in a very difficult position, which arises in part from the particular properties of urban form, which distinguish it from other commodities. Amongst these properties, built form has the characteristic of being 'unavoidable': save for the hermit, everyone has to use it. Unlike the people who produce most other kinds of commodities, therefore, all the members of the development team themselves are also, unavoidably, users of the commodified settings which they are involved in producing. If recent transformations really are widely unloved, then many members of the team must run the ever-present risk of being involved in the production of settings which, as users, they themselves detest – a situation which can all too easily seem senseless indeed.

For a number of reasons, this problem of 'senseless work' is particularly acute for those in the design professions. First, all designers know that each one of them bears some direct, individual responsibility for the production of these unloved transformations. Unlike (say) bricklayers or carpenters, they cannot claim that they are merely 'following orders' in what they do. Instead, they take a 'creative' role in the process – they have a greater degree of autonomy, and have to take personal decisions which make significant impacts on the forms which are finally produced on the ground. Indeed, as we saw in Chapter 2, it is this very element of creativity which gives designers a large part of their job satisfaction. Further, the high degree of creativity which is called for in the design process often encourages designers to see themselves as artists, operating within an art tradition which urges them to draw on their own personal feelings as a key source of raw material in their work. Where their own feelings about the results of their design work are negative ones, as experienced on the ground, the problem of senselessness must be especially hard to keep at bay. Finally, and quite apart from their own personal feelings, designers – as professionals – know that they are meant to work in the public interest, not merely in the cause of maximising their paymasters' profits. Not only are they aware that their own decisions contribute to the production of unloved places, they also know that their professional status entails cultural rules which forbid them to act like this.

For all these reasons, therefore, designers have had to work quite hard, during the capitalist era, to make sense of their working lives, and have constantly been attracted towards ideological material which they could use to construct 'lines of defence' to keep feelings of senselessness at bay. As we saw in Chapter 3, these lines of defence can never just be 'made up' as cynical fictions. They can only work, in helping designers to make sense of their working situations, if they have some foundation in designers' own lived experiences of using the types of places they are led to produce in order to survive in the capitalist market for design services. In practice, therefore, they have somehow to present these experiences, to themselves as much as to others, as positive in users' terms. Constructed in this way, ideological lines of defence form a complex system, as Therborn points out:

> Three successive lines of defence of a given order can be established. First, it can be argued that certain features of this order exist while others do not: for example, affluence, equality, and freedom, but not poverty, exploitation and oppression . . . Second, if this line of defence

no longer holds, and the existence of negative features has to be admitted, it can be argued that what exists is nevertheless just, for example, because the poor and the powerless are misfits and failures who deserve what they get . . . Third, even the existence of injustice may (have to) be admitted, but then it can be argued that a more just order is not possible, or at least not now.[2]

An effective first line of defence has come from emphasising the role of market relationships in the form-production process. Because the types which result from this process are frequently bought in the marketplace (otherwise they would not be in good currency anyway), it is attractive to claim that they must be 'what people want'. As the British planner Peter Hall puts it, for example, people 'vote with their feet and their mortgages' to choose dispersed settlement forms.[3] This 'market' ideology plays a very powerful role in making sense of working to keep the development process going. In principle, it is relevant to all morphological elements, at all physical scales, as well as to all the participants in the development process. As I know from my own experience, it is certainly the principal ideological support of developers themselves, and of the finance-orientated surveyors who are usually their closest advisers.

By itself, however, the 'market' line of defence is not enough to make sense of designers' day-to-day work. Designers – architects, town planners, traffic engineers or whatever – know perfectly well that they have to make their proposals in advance of the approval (or not) of people buying in the market, since that can only be granted or withheld in relation to a finished project on the ground. In any case, designers know from their own experience that most people's interaction with urban places, on a day-to-day basis, does not take place through a 'market' relationship – much of it is inevitably lived out in relation to places which others have paid for.

The problem of 'senseless work', then, creeps all too easily back from under the market carpet; and in recent times many designers have found themselves forming a 'cognitive minority', in the sense defined by the sociologist Peter Berger as 'a group of people whose view of the world differs significantly from the one generally taken for granted in their society. Put differently, a cognitive minority is a group formed around a body of deviant "knowledge".'[4] There is always the potential for doubt about the validity of this deviant knowledge to creep in through the whisperings of the user who always finds a place amongst each designer's own multiple subjectivities. It is difficult to face these doubts, to keep faith with the deviant knowledge, as a lone individual; the support of other deviants is needed to make the sense of deviance disappear:

> Only in a countercommunity of considerable strength does cognitive deviance have a chance to maintain itself. The countercommunity provides continuing therapy against the creeping doubt as to whether,

after all, one may not be wrong and the majority right. To fulfil its function of providing social support for the deviant body of 'knowledge', the countercommunity must provide a strong sense of solidarity among its members (a 'fellowship of the saints' in a world rampant with devils), and it must be quite closed vis-à-vis the outside ('Be not yoked together with unbelievers!'). In sum, it must be a kind of ghetto.[5]

The formation of this counter-community ghetto – 'the compound' as its enemy Tom Wolfe[6] called it – starts early in design education. Its development amongst architecture students is (rather despairingly) charted by Victoria Ellis, in her 1990 commencement address to the students of Illinois University Architecture School:

> Why is it we entered with incredibly diverse backgrounds, interests and friends and we leave here with the exact same handwriting, muttering a language that prevents normal communication and exchange with most anyone outside of our future profession – and we like it this way . . . At parties, the architecture students subconsciously form perfect geometric circles and soon the circles expand to fill all vacant space – so that by the end of your second semester of studio work, your normal (non-architecture) friends will no longer have you.[7]

Though face-to-face contact is important in building this counter-community, it is not enough by itself. To ward off their sense of deviance, many designers need to feel part of some community larger and more broadly-based than their own particular circle of like-minded friends; and it is important also to feel that their (deviant) knowledge has 'proved itself' by lasting over time. Like all of us, no doubt, many designers are therefore drawn towards ideological material which helps them to construct a historical community of people who, they feel, have shared their views.

For design professionals in recent times, the need for this historical community has been filled by the ideology of 'Modernism', or 'the Modern Movement'. Looking back over a hundred years of Modernism from the end of the twentieth century, Weston sees it like this: 'Being modern means being up to date, but being a Modernist is an affirmation of faith in a tradition of the new, which emerged as the creative credo of progressive artists in the early years of the twentieth century.'[8] Modernism, then, has its own 'founding fathers' in the 'progressive artists' of the early twentieth century, such as Frank Lloyd Wright, Le Corbusier and Mies van der Rohe; or Walter Gropius, Tom Wolfe's 'Silver Prince'.[9] It has them, too, in the progressive engineers – quintessential experts – who figure so large in Nikolaus Pevsner's classic *Pioneers of Modern Design*. It has its own tradition in the 'tradition of the new'.

The sense of being part of a Modernist tradition, then, plays a powerful ideological role for many designers, a fact we shall have to remember when we come to thinking how better-loved places might be made. But despite the support which the Modernist community provides, the users' whispers from within can never entirely be stifled. Their nagging drives the more perceptive designers back towards a second line of defence within the walls of Modernism itself: 'OK, maybe users *don't* like what I produce, but it's *good* for them. It may not be what they *want*, but it's what they *need*.' In recent times, claims like these have been strongly underpinned by the ideologies of 'the expert' and 'the artist', whose emergence we explored in Chapter 4.

Expertise offers the underpinning provided by the idea that 'the expert knows best'; inexpert as they are, lay users cannot themselves see their 'real' needs. There is a long tradition of this attitude in the cultures of design, whether in its architecture or its planning strand. As the architect Mies van der Rohe famously put it:

> Never talk to a client about architecture. Talk to him about his children. That is simply good politics. He will not understand what you have to say about architecture most of the time. An architect of ability should be able to tell a client what he wants. He may, of course, have some very curious ideas and I do not mean to say that they are silly ideas. But being untrained in architecture they just cannot know what is possible and what is not possible.[10]

This kind of defence against letting users into the design process has often been strongly reinforced by both the expert's and the artist's emphasis on innovation, whose origins we explored in Chapter 5. From this viewpoint, users cannot know what they 'really' want, because their imaginations are stuck in the limited expectations of their everyday experience: they know what they like, and they like what they know. As the British architect Denys Lasdun sees it, 'our job is to give the client . . . not what he wants but what he never even dreamt he wanted.'[11]

This attitude has formed a very powerful bulwark indeed. It has enabled both experts and artists to believe that they are adopting a generous, positive attitude towards the user, provided that the designer believes that users do not know what they really want. If this belief is maintained, then this line of defence is very hard to breach: even if users stubbornly persist in rejecting what experts give them, then (from the position of this particular ideology) that merely 'proves' that their values should not be taken seriously.

In its most extreme form, the value placed on innovation as positive in its own right – what might be called *neophilia* – enables designers to discount virtually any criticisms which users (or anyone else, for that matter) might make of any new design concept. By their nature, practical criticisms depend on comparisions: *this* is worse than *that*, for example. But the 'that' here must always antedate the 'this', or it would not already be there for comparison: in this sense, it is always transformations which are evaluated. Any argument which rejects as 'nostalgic' or 'backward looking' any negative evaluation of new ideas against the touchstone of older ones, as neophiliacs tend to do, works in practice to protect innovative designers from any comparative, practical critique whatsoever.

Many architects, of course, feel very strongly that they are artists as well as experts. This feeling is not by any means confined to star performers: writing of architects in general, for a newspaper audience, the architectural journalist Hugh Pearman says that, 'He clings to the notion – even when all he's really doing is slipping a skin onto a steel-framed office block – that he is an Artist. This is what keeps him going. This is why he sweated through seven hard years of training.'[12] The culture of 'architecture as an art form' provides a rich vein of ideological material to justify still further the user's exclusion. In principle, it need not make architects deaf to users' reactions. There is, after all, a range of art forms in which audience involvement is an important part of artistic creation itself: actors in the live theatre, or musicians making certain kinds of live music are cases in point here. The particular form of art which has typically been called on in defining the architecture of recent times, however, is not like that. Instead, architecture has been defined as a fine art, in the mould of painting or sculpture. This romantic fine art view, dramatised in the *Fountainhead* syndrome,[13] focuses only on the individual architect's creativity. It has become deeply embedded as a constitutive rule in the way many architects themselves conceive of their proper role; and if we take a trip to the library we find that it is currently institutionalised in the way knowledge is socially stored too: the architecture section is near to sculpture and painting, but far from theatre or music or any other form of art which might involve the user or the audience in the process of artistic creation itself.

To anyone who accepts this romantic, fine art definition (and this includes many users as well as architects themselves) it is just 'obvious' that users cannot be thought of as making creative contributions to the art of architecture. They merely constrain what architects can do, for architectural creativity comes from architects alone. When asked 'should there be a dialogue between the public and the architect?', for example, the German architect Oswald Matthias Ungers illustrates this central tenet of so much current architectural culture with an answer which would doubtless be echoed by many fellow professionals, had they his bottle in these times of political correctness:

> There can't be. The process of being creative – I'm a little bit hesitant to use this word but I have to do so – is a dialogue between you and your fantasy, your intellect, your ratio, your mind and your potential. That is the dialogue.[14]

This means (so the argument goes) that if users' views *are* taken seriously, then it is all too likely that bad design will result. The Austrian architect Hans Hollein expresses this feeling in the most uncompromising form:

> Architecture is not satisfaction of the needs of the mediocre, is not environment for the petty happiness of the masses. Architecture is made by those who stand on the highest level of culture and civilisation, in the vanguard of the development of their era. Architecture is the concern of elites.[15]

This was written back in 1962, but the attitude lives on. The American critic, Bruce Goodwin, makes the same claim, more recently, in a less pretentious way: 'Now the architectural consumer wallows in butter-whipped mediocrity. The vitality of art is possible only in confrontation and opposition to middle class values.'[16] The definition of architecture as a romantic fine art also has implications for the kinds of people who are attracted into the profession in the first place. Studies of architecture students, for example, strongly suggest that architecture attracts a particularly high proportion of 'introverted intuitives', who tend to develop subjectivities which are not particularly well-attuned to the involvement of others in the creative process.[17]

There is also an important gender dimension to the cult of artistic genius which underpins the romantic fine art tradition. As the philosopher Christine Battersby shows, there is a long tradition of Western thought which sees genius specifically as a male attribute,[18] and therefore prevents women from being taken seriously as great architects. In popular culture too, women's creative efforts have largely been pushed into the private realm, to engage in their 'natural' creative act of producing babies, and the art of home-making – a move which further supports the maximisation of commodity consumption in the twentieth century's dispersed settlements. Here is an advert for baby-food, from one of the spate of home-orientated magazines which gave such powerful ideological support to this dispersion during the 1920s:

> Do you realise, mother of a baby – that you can be one of the greatest creative artists in the world? – Just like every other artist a mother must study her art – the greatest art of all – the great art of creating strong, straight, noble men and women.[19]

This view that the 'natural' art for a woman involves the creation of babies rather than artifacts is current not only in popular culture but also in the elite culture of fine art itself. Here is the British sculptor Reg Butler putting this point of view to a student audience:

I am quite sure that the vitality of many female students derives from frustrated maternity, and most of these, on finding the opportunity to settle down and produce children, will no longer experience the passionate discontent sufficient to drive them constantly towards the labours of creation in other ways. Can a woman become a vital creative artist without ceasing to be a woman except for the purposes of a census?[20]

It is 'unnatural', then, for women to be artists. This is not just Butler's own idiosyncratic point of view: as Roszika Parker and Griselda Pollock point out, its widespread current acceptance in the worlds of art production and art criticism is manifest in the gendered way in which language is used in those worlds: 'The reverential term "Old Master" has no meaningful equivalent; when cast in its feminine form, "Old Mistress", the connotation is altogether different, to say the least.'[21] The fact that relatively few women are taken seriously enough, as artists, to acquire large amounts of cultural capital reinforces architectural culture's user-deafness. Almost everywhere, women are indeed born into a world where it is largely they who are expected to take responsibility for helpless infants who would otherwise die. This system can only reproduce itself if many women do in fact develop 'feminine' subjectivities in which rules about nurturing and caring for others play important roles. The virtual exclusion of the very people most likely to hold such user-positive values from positions of power and influence in the architectural world effectively robs architectural culture of elements which might otherwise make it more responsive to the user's voice.

The ideologies of 'the expert' and 'the artist', sharing the concept that 'good design' exists in some absolute sense, also enable designers to work with clear consciences in the global arena within which they increasingly have to market their design services. Clearly no individual designer can ever hope to acquire any deep understanding of all the varied cultural situations within which so many of them nowadays have to work. The idea that there is some absolute touchstone of 'good design', floating free from ties to any particular culture, offers a powerful defence against the debilitating doubts which might all too easily arise in this situation.

In their current forms, then, the ideologies of 'the expert' and 'the artist' both provide powerful general defences for what designers do. But designers have to make myriad day-to-day decisions about the particularities of design. To form effective lines of defence, therefore, the generalised conceptions of the expert and the artist have to be backed up by more detailed ideologies, which support the ways in which each element of the typological repertoire is typically transformed in the quest for profit.

Transformations at the largest scale, from compact to dispersed settlement patterns, were initially supported by reference to issues of public health. Many of the industrial cities of early capitalism were the sites of life-threatening diseases such as cholera; and it was easy to see why people would move away from the worst-affected areas – the densest areas, inhabited by the poorest people – for their own protection if they could afford to do so. These moves were incorporated into design culture, at a professional level, by all sorts of innovative settlement ideas which proposed to make cities more healthy by giving their inhabitants easier access to the benign qualities of 'nature'. For example, the 'Garden City' concept proposed by Ebenezer Howard,[22] with its search for a harmonious equilibrium between town and country, was hugely and internationally influential. This influence, however, operated in rather an oblique way. Significantly, very little was actually built according to Howard's precepts; in practice, the book proved influential as a justification for suburban estate development of a type which Howard himself deplored. As we should expect, the aspects of Howard's ideas which proved saleable in the marketplace for urban innovation were those which supported capital accumulation, whilst his other ideas, including an emphasis on social facilities, were largely ignored. Nonetheless, part of the saleability of the suburbs themselves certainly depended on the 'health through contact with nature' idea, which found powerful expression and reinforcement during the inter-war years in endless advertisements for housing estates incorporating images of newly-arrived suburbanites throwing open the casements to savour their new fresh air to the full.

Though design professionals often looked down their noses at the suburbs – possibly because these were often developed without the involvement of professional architects, and therefore offered few employment opportunities[23] – even the most avant-garde of architects were in favour of more 'architectural' versions of what Le Corbusier, in sloganising mood, called 'Sun, Space and Green' as key principles of good design. From the 1970s, however, these links between dispersal and health were increasingly called into question. Both radical and establishment voices, from Friends of the Earth to the Council for the Preservation of Rural England, began promoting the idea that the expansion of low-density cities was now a threat to the countryside itself, so that new, less dispersed settlement forms should be sought. These worries were paralleled by increasing levels of concern about the pollution caused by the high levels of motorised transport which dispersal itself had promoted. 'Health through Nature', unaided, was no longer enough to make sense of the transformation towards more dispersed settlement patterns. New ideological support was needed.

From the welter of ideas available in marginal areas of design culture, this demand brought into prominence two new, related strands of ideological material with potential to support the dispersed settlement. First the old term 'suburb' which, as many have pointed out over the years, implied that dispersal was a mere 'add on' to some 'proper' concept of a town, began to be edged out in professional parlance by the more 'positive' term 'Edge City', with its connotations of active notions like 'the cutting edge'. Second, a long-running critique of suburban living as dull and conformist began to be challenged by the concept of Edge City as the 'new frontier'. Portrayed now as lively, interesting and where the action is, this image is perhaps articulated most memorably in the book *Edge City* by the American journalist Joel Garreau.[24]

The second of the key transformations of overall settlement patterns which we explored in the last two chapters – the shift of land use patterns from a fine grain of mixed uses towards larger, mono-use 'zones' – finds ideological support within design culture from the argument that grouping like with like avoids conflicts between 'incompatible' uses. This is a powerfully persuasive argument, which may or may not be true in particular circumstances. Evidence suggests, however, that it probably came to prominence in planning culture at least as much because it helps planners to rationalise what they have to accept, in doing the bidding of current power-holders, as because it positively helps them to plan. This is strongly suggested by studies of how zoning policies work out in practice. William Stull, for example, in a study of zoning in practice in Boston, Massachussetts, found that

> Zoning laws are ineffectual in actually constraining the location of economic activities . . . [as] it often appears that land market forces are too strong for officials administering a zoning statute to ignore. Zoning regulation is constantly being adapted to accommodate these forces. If there is time, the long run locational equilibrium in a zoned community will consist of a configuration of land uses which is not very different from that which would have occurred had zoning never been introduced in the first place.[25]

Policies of zoning, then, seem effective in confirming and legitimating the status quo, rather than in changing it. In the process, though, they certainly help planners to make sense of what they have to do in the practice of their work. This being so, there are probably few practising planners who would agree with Stull when he goes on to argue that 'under such circumstances, it may make sense to eliminate zoning laws and save the costs associated with their administration.'[26]

Within the transformations of the overall settlement pattern whose ideological supports we have discussed so far, the capitalist era has seen the transformation of public space networks from grids to hierarchies, with a proliferation of enclaves. This shift has received powerful ideological support from two key themes: safety and community.

The safety theme initially grew from an increasing public concern for road safety which emerged, first in the United States, as car-ownership rates rose during the 1920s. In response to this concern, for example, the American planner Clarence Stein embodied innovative ideas for placing homes around car-free inner courts in a seminal housing scheme at Radburn, New

Jersey (Figure 6.1). The renewed ideological support which this gave to the enclave system became typified as the 'Radburn principle', and was to retain its ideological power for half a century and more.

This idea of separating different modes of traffic was amplified and broadcast more widely during the 1930s. In his highly influential *Road Traffic and its Control* of 1938, Alker Tripp developed a sophisticated system for segregating different modes through the classification of roads – a thorough-going application of 'functionalist' design principles to public space, which had previously been conceived of as being for multiple users together. The classification of roads into qualitatively different types – main roads where pedestrians, cyclists and parked cars were excluded, as distinct from minor roads which excluded all motorised traffic except for local access – amounted to a constitutive rule which fragmented the basis on which public space was conceived by its designers. Encouraged to go fast on the main roads, people driving motorised vehicles – now conceived as depersonalised and depoliticised 'traffic' rather than as people with particular interests – made intolerable impacts on others living or working nearby. This gave further justification for enclave subdivisions, an argument persuasively made by Colin Buchanan in his influential *Traffic in Towns* of 1964. Here, the city becomes divided into 'environmental areas'

Figure 6.1 The Radburn layout.

which are each kept free from through-traffic, an attitude which is still widely current today.

As levels of comparative deprivation and social stress began to rise in many 'advanced' capitalist countries from the 1970s onwards, further ideological support for the move towards enclaves began to be drawn from ideas about 'defensible space' put forward by the American architect Oscar Newman.[27] Newman proposed that levels of crime might be reduced by grouping buildings, particularly housing, around culs-de-sac onto which they would all contribute the maximum surveillance. This idea was taken up in Britain by the geographer Alice Coleman, whose *Utopia on Trial* showed how it could be related to conditions on run-down British housing estates.[28] In turn, Coleman's work was taken up, in Britain, both by the right-wing Thatcher administration and by the police: it gave rise, for example, to the police campaign entitled *Secure by Design*,[29] which gave strong support to the enclave principle by proposing that new housing areas should be built with only one entrance from the outside world. A further twist to this 'retreat to the enclave' is provided in the gated, guarded 'communities' which, beginning in the United States, have now become an international type.

The fragmentation of public space into enclaves has also received powerful ideological support through the theme of 'community'. From the early days of capitalist transformations, commentators of the left have criticised the enclave tendency for its socially divisive effects: recall, for example, Friedrich Engels's complaints about Manchester's 'hypocritical plan', (see p. 101). Faced with such criticisms, the most powerful ideological riposte has been to turn the criticism on its head, claiming that the proliferation of enclaves in fact supports the development of a sense of community. For example, the British architect Augustus Pugin, writing about the same time as Engels but from a viewpoint as a fervent Catholic romantic, with politics a million miles from Engels's atheist Marxism, wrote and drew a brilliant polemic entitled *Contrasts*. Evocative drawings of introverted spatial forms, such as the cloisters and quadrangles typical of medieval religious buildings, were juxtaposed to captions emphasising the supposedly humane, community-orientated values of the Middle Ages in contrast (hence the book's title) to the abstract inhumanities of the contemporary capitalist city.

The idea that introverted spatial forms somehow create 'community' (however defined) has remained a commonplace of mainstream architectural ideology ever since. It underlay the design of the nineteenth-century philanthropic estates,[30] looms large in Le Corbusier's *Ville Contemporaine*,[31] and *Plan Voisin* of the 1920s – the latter, significantly, sponsored by an automobile manufacturer – and as late as 1977 formed the basis of 'good housing layout' so far as the Greater London Council was concerned.[32] The architectural theorist Bill Hillier sums it up like this:

In the recent past, a common social value in housing design has been that of the small, relatively bounded community, forming an identifiable unit of a larger whole. Architecturally this has been reflected in a preoccupation with linking groups of dwellings to identifiable and distinct external spaces in the hope that the 'enclosures' or 'clusters' so created would help group identification and interaction.[33]

Further support for the shift from grids to enclaves has come from aesthetic quarters. Reacting against the supposedly monotonous and boring character of the regular grid-plans of much nineteenth-century speculative development, theoreticians such as the Austrian architect Camillo Sitte, in his *City Planning According to Artistic Principles*,[34] put forward a highly selective reading of the pre-capitalist city. Stressing the spatial enclosure of the various parts of the medieval public space system, rather than the integrated continuity which was in fact typical of the system as a whole, and which we shall explore later, Sitte's reading gave strong aesthetic support and historical legitimacy to the enclave approach (Figure 6.2). Much later, Gordon Cullen's evocative *Townscape*[35] illustrations (Figure 6.3) reinforced the same point, in the process making a bridge between architecture and town planning. Of all 'design' books, *Townscape* is the most widely-read in planning circles, in Britain at least.

Figure 6.2 Enclosure as a principle of urban design: Piazza Erbe, Verona.

Figure 6.3 An enclosed 'townscape' space.

At the smaller scale of the particular building project, ideological support for the shift from connected, terraced masses to free-standing pavilions has been constructed through twin themes of health and art, through the powerful conflation of the 'sun, space and green' ideology with the art concept of 'flowing space', by key architectural figures such as Le Corbusier during the 1920s. Intellectual support for this concept was contributed by claims that 'flowing' spatial arrangements represent the 'spirit of the age': by the 1940s, for example, Siegfried Giedeon had constructed a full-blown history of the 'flowing space' concept, treating it as an important aspect of the character of modernity as a whole, in his *Space, Time and Architecture*. First published in 1941, this went through sixteen editions to become the bible of a whole generation of architecture students throughout the English-speaking world.

According to Giedeon, this new spatial order did not relate only to indoor spaces. One of its key characteristics was what he called the 'interpenetration of outer and inner space', exemplified, for example, in the Eiffel Tower: 'The stairways in the upper levels of the Eiffel Tower are among the earliest architectural expression of the continuous interpenetration of outer and inner space.'[36] This notion of 'interpenetration' acted as a tacit ideology to reinforce the transformation from connected masses of buildings into free-standing pavilions, which has been so characteristic of capitalist development. If 'modern' space flowed (or ought to flow) seamlessly from inside to outside on all sides, which was what Giedeon implied, buildings were (or should be) free-standing pavilions, as the Eiffel Tower was, with nothing solid alongside to break up this spatial flow.

Space is here presented as a desocialised (and therefore depoliticised) medium of artistic expression. It is a concept which leaves no space for socially-constructed notions like 'front' or 'back' – a cultural shift which is emphasised by the denigration of such notions, and of architectural concepts like 'the façade' by which they were made manifest in concrete form. Giedeon himself, for instance, sees nineteenth-century façades 'merely as curtains, disguising what was behind them'.[37]

Further tacit support for pavilionisation is built into the current process of architectural education. The current practice in almost all schools of architecture is to start new students on designing individual small buildings on sites which have little impact on the public realm, or even on designing objects like furniture, which need have no relationship with the public realm at all. This encourages students, at the most impressionable stage in their design education, to conceive of architecture as the design of discrete objects, rather than the making of interventions in larger built form systems. This initial focus on 'the object' is firmly backed up by the current drawing conventions of architectural culture. Here plans and elevations – much the most common drawing types – are usually drawn with their surroundings left out altogether, or else only indicated in schematic ways. Every time they make such drawings, or see them published in the professional media, architects 'remind themselves' that buildings are free-standing objects.

The further transformation from resilient, general purpose building types to the current range of specialised use-types has been supported, at a tacit level, by constitutive rules embedded in a range of classification systems which are taken for granted in everyday working life. For example, the meaning of the phrase 'building type' itself has gradually shrunk, in everyday practical usage, from the connotations of combined form and use, which it had when it came into good currency in pre-capitalist France, to its current connotations of use alone. This shrunken usage is now embedded in library classification systems, and in historical writing: the structure of Nikolaus Pevsner's standard *History of Building Types*, for example, is entirely organised around sections which deal with buildings intended for particular uses.[38]

When designers call on these types in the design of particular buildings, this built-in specialisation is taken a step further through the rules embedded in the functionalist theory of design, whose importance in mainstream architectural culture we have already remarked. Here the most

important determinant of design is usually taken to be the *internal* convenience of the building's various spaces, from which all else follows – 'the
plan is the generator' as Le Corbusier put it in the 1920s.[39] To designers
who take this line, the exterior form of the building – and therefore the
building's relationship to public space and public experience – is merely a
by-product of the internal planning. The character of this approach comes
over very clearly in this more recent account by the New Zealand architect
R.J. McMillan:

> The plan becomes the determination of everything from the very start.
> It implies the methods and construction to be used; starting from
> within to without, for a house or hotel is an organism comparable to
> a human being.
>   The plan proceeding from within to without can be likened to a
> bubble. This bubble is perfect and harmonious if the pressures inter-
> nally are regulated. The exterior is the result of the interior.[40]

At the tacit level of practical technique, this approach is exactly paralleled
by the widespread use of 'bubble-diagram' techniques for the planning of
buildings. Here designers start off with a list of the internal spaces agreed
with the developer, consider the degree of 'functional' linkage between
these, and then relate them together so as to maximise the efficiency of
the linkage system overall, formalising the result into a building plan by
wrapping walls around the resulting activity-spaces. On the face of it this
appears to be a purely pragmatic, technical approach to design. In practice,
though, a moment's thought is enough to see that it considers design only
from the point of view of the convenience of the particular building: the
building which forms the only *saleable* element in all this. This is of vital
importance to the developer, but the interests of the wider public, who
mostly experience any particular building only in terms of its impact on the
public realm, are left out of account altogether.

At the smallest scale of the typological repertoire, the transformation of
building surfaces from local to global in terms of their construction materials, and from complex to simple in visual terms, draw support from a range
of ideological sources. To start with, the financial pressure to reduce the
number and scale of design drawings made it increasingly difficult for
designers to focus on small-scale surface detailing in any case: a 1:100
drawing, looked at from a half-metre distance, can only simulate the
experience of a building which is fifty metres away.

From the beginning, however, these surface transformations were not
merely supported by such tacit ideological material – they quickly became
too important to too many designers for that to suffice. They became
important because they fulfilled the need for a distinctive formal vocabulary
through which members of the Modernist counter-community could give
external expression to their mutual solidarity. Opportunities for developing

such expressions were limited. On the one hand, they had to be highly visible, and on the other they had to be able to be seen as in some way consistent over the timespan of the community itself. Taken together, these twin imperatives could only be fulfilled at the smallest scale of the typological repertoire; as we have just seen there are powerful pressures for those at the largest scales to undergo radical transformations during the twentieth century. These pressures could be resisted relatively easily at the scale of the surface elements, which have such an important impact on the 'look' or 'image' of a place, but are relatively independent of transformations of elements at the larger scales.

Modernist designers were therefore drawn towards ideological material which supported an aesthetic of simple surfaces, through claims from the founding fathers and later heroes of Modernism that building surfaces and details *ought* to be visually simple. This argument that 'simple is good' runs as a leitmotif through much modern architectural writing, from Adolf Loos's insistence that 'ornament is crime', with its maxim that 'the evolution of culture marches with the elimination of ornament from useful objects',[41] through Mies van der Rohe's celebrated aphorism that 'less is more'. The transformations of building surfaces from local to global, from complex to simple, from figurative to abstract therefore became the enduring trademark of the Modernist counter-community – an aesthetic trademark, whose subjective significance throughout that community cannot be overstated.

This emphasis on the historical continuity of a specifically Modernist aesthetic is not to deny the fact that many designers, seeking to develop their own 'unique selling points' to compete in the marketplace, have offered other 'looks', particularly taking advantage of the general cultural space opened up by 'postmodern' upheavals. Nor is it to deny that some of these alternatives have been taken up enthusiastically by developers from time to time, sometimes with evident public approval. Interesting though these alternatives are, however, it seems to me that they can only realistically be seen as exceptions which prove the general rule that the Modernist aesthetic has tremendous staying power within design culture itself, as the visual affirmation of the designer's counter-community, and as an affirmation of the importance of the visual, aesthetic dimension in keeping this supportive sense of community in being.

In this connection, it is surely significant that postmodernism has had a very particular kind of impact on architectural culture – an aesthetic impact, as Bernard Tschumi points out:

> It is striking to notice, for example, the respective interpretations of postmodernism in the separate fields of art and architecture, whereby postmodernism in architecture became associated with an identifiable style, while in art it meant a critical practice.[42]

Even the many attacks which have been mounted against this identifiable postmodern style have revealed the importance of 'the look' in architectural debate:

> Polychrome brickwork, brightly painted planks of wood, rustic iron-work, gables and slate tiles became the clichés of the 1970s . . . the new shopping complexes, civic centres and housing estates looked as though they had come, jaded, from an all-night fancy-dress ball.[43]

By focusing almost exclusively on the aesthetic level, even furious attacks like this – which have been such a noticeable feature of recent architectural debate – have themselves emphasised the importance of the aesthetic dimension in the broadly modernist position from which they have usually been launched. Indeed, by supporting modernists in taking this overtly aesthetic stance, postmodernism has paradoxically taken its revenge. Its eventual impact on the Modernist aesthetic seems to have been to rescue it from postmodern questionings, but only as a style to be revived. As Robert Venturi comments, with rather puritanical disapproval, 'An ultimate oxymoron: Neo-Modern?'.[44]

The Modernist aesthetic, then, is in practice as central to the mainstream design community now as ever it was; it is now espoused, as Venturi himself acknowledges, with greater enthusiasm than ever: 'In their fervour to uphold Modern – its style, not its principles – against historical reference neo-modernists' vehemence is almost menacing.'[45] And, as we should expect from the globalising tendencies of the development process itself, this resilience of the Modernist aesthetic is itself a global one; as Weston points out: 'If the example of Japan is any guide, the aesthetic and technical resources of Modernism will prove a continuing source of inspiration in other rapidly developing Asian countries.'[46]

To summarise, we can now see that a wide range of ideological material has developed in recent times, to support as 'good design' the transformations of the typological repertoire at all scales which have evolved, through the marketplace, to serve the interests of capital accumulation. We have also seen, however, that the effectiveness of these 'good design' concepts has increasingly come under attack since the 1970s. Designers have progressively been driven back to a final line of defence of what they do: 'maybe users *don't* like recent transformations, and maybe they are *not* good design, but none of this is *my* fault.'

Taking this line, professionals from all the specialised spheres of the development process can avoid feeling personally responsible for their own involvement in the making of unloved places. Within the division of labour of the complex 'development teams' which characterise the current development process, town planning (for example) is concerned with setting a context which both enables and constrains the architecture of particular projects. Given the practical overlap between town planning and architecture, it is hard to understand the logic of their separation in practical terms. In ideological terms, however, it helps practitioners in both fields to feel justified in ignoring users' criticisms of the places in whose production both are involved, for it creates the potential for what the Spanish architect Manuel Sola Morales calls 'the great alibi', through which those in each camp can blame the other for any failings which users point out.[47] The great alibi, of course, encompasses all the various actors in the fragmented 'development team', the opportunities for evasion which it offers are taken up with enthusiasm, and blame is shovelled around with gusto, through the mutual antipathy between the team's various 'expert' and 'artist' members which we have already remarked, and of which the economist Ralph Morton reminds us:

> The surveyor's or engineer's view of the architect is often very different and far less flattering than the architect's view of himself. The architect's view of the structural engineer is often as someone who is highly intelligent and useful but essentially philistine (since first writing this sentence I have heard a civil engineer refer to quantity surveying as a 'parasitical profession'). As for views expressed, by architects, of building surveyors, they are best left unquoted.[48]

The upshot of all this is that very few people (except perhaps the developer) currently have a detailed overview of how the development process works as a whole. And this means, of course, that no one is in a strong position to challenge whatever the developer wants done. Given the difficulty of controlling complex professional work, this is just as well for the developer. No more effective strategy of 'divide and rule' could have been dreamed up by the Iron Chancellor Bismarck himself.

Arriving at this final line of defence, we have completed our review of those aspects of the current cultures of design which enable designers to make sense of carrying on their work in today's development process. In terms of the constitutive rules which define what a designer is, we have seen the central importance of particular conceptions of 'the expert' and 'the artist'. We have seen the negative effects these both have in excluding users' voices, and we have seen the cumulatively negative impact of their mutual antagonism, in contributing to the 'great alibi' which gives a last ditch of ideological support to carrying on as things are. Overall, we have also seen how the evaluative rules which define 'good design' for many mainstream designers end up supporting developers' interests in the accumulation of capital, by calling on the key themes of choice, community, safety, nature, health and art.

Unless we are to believe that all this is just a complex set of coincidences – a supposition which seems too far-fetched to be credible – this gives solid support to the idea that many current canons of good design are not at all the unpolitical affairs they are made out to be. In their current forms, they are better seen as formidable weapons deployed to shield the power bloc's interests. But these defences, no matter how sophisticated and resilient they may be, are nonetheless far from impregnable. Despite all the limitations of the current mainstream cultures of design, designers themselves are not robots – they are human beings who always have some freedom to think new thoughts and to feel in new ways. Designers who want to get beyond the status quo, to heed the voice of the user within and to work towards better-loved places, will have to find less defensive, more creative ways of putting this freedom to use. Before we go on to seek these new ways in practice, however, let us consolidate what we have come to understand through the last three chapters as a whole.

Conclusion: supports for the power bloc

In the last three chapters, we have used our structurationist frame of reference to explore the transformations which the form-production process and its physical and ideological products have undergone during the capitalist era. Through this exploration, we have identified a multiplicity of ways in which these transformations have supported both the process of capital accumulation itself, and the maintenance and reproduction of the social and economic arrangements which have made high levels of capital accumulation possible in the first place. Before we go any further, let us weave more firmly together the multiple threads we have teased out so far.

First, we have come to understand how the working of the capital accumulation cycle has supported the development of an ever more complex set of divisions within the overall body of those involved in producing and using urban settings. At the most fundamental level, we have seen how the making of profits can only occur through the creation of a consumer–producer gap – a separation of the producer from the consumer who buys the commodified product. On the production side of this gap, we have seen how the spread of market relationships has fostered the growth of a wide array of professions, operating in an increasingly global marketplace, fragmented through their development of saleable 'unique selling propositions', supported on the one hand by the ideology of 'the expert' and the 'artist', and on the other hand by the progressive deskilling of most other workers to form a cheap, easily-disciplined workforce. At the level of the urban product itself, we have come to see that the working of the capital accumulation cycle has fostered transformations in all the morphological elements with which we are concerned, together with canons of good design which offer these transformations ideological support. Through the review we carried out, we have come to understand how forms and ideologies alike have developed as highly effective power-bloc supports.

Considering first the largest scale of typological transformations, we have seen that the shift from relatively compact to relatively dispersed, lower-density settlement patterns helps to keep the capital-accumulation process going in the face of persistent problems of under-consumption, through the way in which it requires people to increase their consumption of all sorts of commodities, from transport to freezers, if they are to maintain access to any given range of human contacts, goods and services. The central role which this transformation towards lower densities plays in propping up the system is supported by ideologies which are shared within both popular and professional cultures, through appeal to the linked concepts of health and nature.

The way in which public space networks have been transformed, both as large-scale systems and in their particular constituent spaces, also supports both the direct making of profits and the reproduction of the overall system which makes profit-making possible. Most obviously, at the larger scale, the transformation from highly-connected grids to more hierarchical tree-like systems adapted for faster vehicular movement helps to increase the profits of many enterprises by speeding up both the large-scale long-distance movement of raw materials and labour, and the movement of completed products to their points of sale. Perhaps less obviously, this transformation – supported by the ideology of 'functional efficiency' – helps to keep the overall system going by splitting areas of settlements off from each other on 'opposite sides of the tracks', thereby reducing everyone's awareness of social deprivations, injustices and tensions – a support for the power bloc which is reinforced by the smaller-scale fragmentation entailed by the parallel shift from connected streets to separated enclaves. Paradoxically supported by the ideology of 'community' which pervades both popular and professional cultures, the proliferation of enclaves also increases the potential profits of property investment by creating a 'unique selling proposition' which is exclusive to the enclave itself, and by buffering particular developments within the enclave against future uncertainties from transformations outside it.

We have also seen how support for powerful interests within the power bloc is offered by transformations in patterns of land use. The shift from a relatively fine grain of mixed uses towards a far coarser, more separated-out pattern enables landowners, for example, to maximise the economic value of their holdings by accommodating only the 'highest and best use' which, in the particular location concerned, generates the highest land value. In terms of propping up the system, the spacing-out of different uses into separate areas – supported in design culture by the ideology of 'zoning' – provides a further level of dispersion, and generates still further increases in the consumption of all sorts of commodities associated with the amount of vehicular traffic needed for people to move between the resulting mono-functional zones in the course of living normal life. At the smaller scale of the particular development site, we have seen how the shift from relatively small plots to relatively large ones reduces developers' costs through econo-mies of scale, and therefore increases their competitiveness in the market-place. Since larger plots allow greater freedom to arrange buildings and open spaces within the plot itself, the move towards larger plots – supported through the ideology of 'comprehensive' development or 'total design' –

also offers more opportunities for the shift towards enclaves, and for the development of new profit-maximising building types.

The transformations undergone by building types themselves, and by the typical relationships, both between particular buildings and between buildings and public space, have also evolved in ways which provide increased support for the power bloc. In particular, we explored the shift from robust 'loose-fit' buildings to an ever-wider range of more specialised types, adapted to particular patterns of use, supported by the ideology of 'function'; and we saw how this change helps to prop up the system by making it easier for those with power to control the patterns of activity which go on inside the buildings concerned, at the same time as it increases developers' profits by reducing the floor areas within which particular activities can happen, and therefore reducing the producer's rental outgoings.

In terms of the relationships between one building and another, we have seen a transformation from connected masses towards free-standing pavilions, supported by the ideology of 'design in the round', as developers sought to give each of their products its own unique selling proposition in the speculative marketplace. In parallel, in terms of keeping the overall system running in the face of its endemic problems of under-consumption, we have seen how these transformations towards a proliferation of separated buildings have helped to increase the consumption both of building materials and of the energy needed to keep the buildings heated and cooled.

Finally, in relation to the interface between buildings and public space, we have explored a range of transformations which flow from the ways in which built form has come to provide power-bloc supports. For example, the search for the most cost-effective materials in a global market, calling on the advantages of cheap production labour wherever available, has led to a dual effect. On the one hand, it has led to a transformation from local to global materials, but on the other hand the desire for simple purchasing processes and straightforward organisation of the construction process has led to a situation, supported by the ideology of 'simplicity', in which only a limited palette of materials is used in any one building. The cumulative upshot is that everywhere has become more and more varied in its appearance, just like everywhere else.

The shift towards free-standing pavilions has also had a radical impact on the walls of the public realm, since it has led to many situations in which private 'backs' face on to public space, with a consequent shift from an interface formed primarily of 'active' fronts towards an ever more 'passive' character. This has often been compounded, we saw, by a radical reduction in the repertoire of types of surface details, aiding the process of deskilling construction workers and the move towards repetitive assembly techniques – all again supported by the aesthetic ideology of 'simplicity'.

By this point, we have seen that many recent transformations in urban form and in design culture can be explained by reference to the roles they play in the capital accumulation process. We must always remember, however, that our exploration has been conducted at a typological level: we have reviewed the kinds of transformations which have *typically* happened. There is nothing inevitable about what has happened in any particular case: none of the patterns we have revealed has the status of an inexorable 'law', determining what happens in every instance. In the end, the form-production process depends on the actions of the myriad human beings who make it run, and all of them have some irreducible degree of autonomy, depending on the resources they can deploy to make things happen. In the end, as we have seen, typological changes come about through a process of negotiation and struggle through the typification process. If we want to work towards the making of better-loved places, we have to learn how to win these battles. We have to find out how each of us can best focus our limited autonomy towards the production of better-loved places, in the light of what we now know about the pressures which constrain our actions in the form-production process. That is the subject of the remainder of this book.

Part III

Positive values, negative outcomes

Introduction

Our explorations so far have revealed the logic with which urban settings have been transformed to support the process of capital accumulation in the face of the internal contradictions which bedevil it, and have shown us the powerful ways in which ideological developments have promoted these spatial transformations. In recent years, however, the system has begun to encounter new limits, with fundamental conflicts beginning to surface within the power bloc as a result of fears about the negative ecological impacts of recent transformations.

These conflicts find expression at both political and economic levels. At the political level, for instance, there is widespread concern about the public health implications of these ecological impacts. Perhaps more salient for change, serious cracks are beginning to open up at the economic level. For example, the major insurance companies – themselves massive funders of built form as long-term investments – are beginning to experience serious damage from huge insurance payouts in relation to 'natural' disasters which are increasingly seen as flowing from the ecological impacts of urban development. Situations like this create major conflicts of interest – for example between the insurance industry, whose already formidable power and influence seems set to increase as State welfare provisions are reduced in many countries, and the oil and transport lobby whose profits are strongly supported by recent urban transformations.

All in all, then, the power bloc has to address serious new problems to keep the system going. These problems, however, are currently ignored by most of those involved in the form-production process itself. Each developer or investor, for example, has an economic interest in being what Jowell and others call a 'free rider':

> Any individual's contribution to both the problem and the solution is self evidently small, yet the costs to an individual of changing his or her behaviour may be large. The truly rational (self interested) individual will want to be a 'free rider', leaving other people to foot the bill while making none of the sacrifices that environmental improvements will require.[1]

In conditions of increasing regional and international economic competition, the State agencies of any particular locality also want to be 'free riders' on the backs of rival regions; and this means that effective political attempts to address the ecological issues posed by recent urban transformations will have to be mounted on the international stage. Political and economic conflicts within the power bloc therefore create a demand for political action on a global scale, attempting to ensure the continued viability of the capitalist system as a whole. This political demand is beginning to be realised in practice, through the growing international influence of the 'sustainability lobby'. Events such as the Rio Earth Summit of 1991, the so-called Agenda 21 to which it gave rise, the European Community's Fifth Environmental Action programme of 1993, the 1998 Tokyo Summit and the World Bank's Global Environmental Facility together represent serious attempts to build a potentially powerful supra-national movement, composed of State and non-governmental agencies, attempting to intervene in the production of urban form to address what is increasingly seen as a global ecological crisis. In practice, however, the political power of this environmental movement depends on its capacity to achieve the widespread public support which it needs if it is to push change through in the face of local economic opposition. This need for mass support creates a new opportunity-space, offering new potentials for political action to push for new types of settings – more lovable as well as more ecologically benign in terms of factors such as climate change.

There are opportunities, then, for getting better-loved places on the ground, but these opportunities will not generate better places through some automatic process. Rather, better places will be made only through the efforts of people who actively desire them. Though political action may create a situation where it is possible for developers to tolerate user-positive transformations without economic demise, developers themselves cannot be expected to demand changes which are unlikely to offer short-run economic benefits. The alliance between users and the environmental lobby is therefore not sufficient in itself to get better-loved places on the ground: further allies are needed, within the development process itself. To some extent at least, all the members of the development team form potential allies here, simply because all producers are also users themselves, necessarily living out their off-duty lives within the types of settings which they themselves are engaged in producing, and therefore experiencing any negative impacts these settings might have on the course of their own everyday lives. For a number of reasons, however, it is designers who probably offer the best alliance-potential in practice.

Despite all the user-negative ideological supports which we explored in the last chapter, designers still find themselves in a special situation which makes it particularly difficult for them to make total sense of any user–producer tensions which might arise within their own subjectivities. As we saw in the last chapter, designers make sense of their work in part by considering themselves as artists, within a particular tradition of art which encourages them to call on their own subjective feelings as sources of inspiration for their work. As important creative wellsprings, these feelings can never be completely rationalised through professional ideologies; if they were, designers would run the risk of losing the very capacity for artistic innovation which is central to their survival in the market for design services. To the extent that designers' own personal experiences as users of given settings might in part be negative, and to the extent that these negative feelings cannot completely be rationalised through ideological means, there must always exist at least some basis for building a user–designer alliance.

This potential is enhanced through the ideology of professionalism itself. Though the paid work they do aligns them in the producer's camp, we have always to remember that design professionals gain important advantages of professional status because of their institutionalised claim to serve some wider public interest in their work. To put it crudely, they know that they are not supposed to go round working against wider users' interests for their own financial gain. This opens up the potential that they might at least try to use the limited degree of freedom which they have in the development team, which derives from the difficulty which developers find in controlling complex design work, to push as far as possible towards more user-positive settings.

It is extremely important to find ways for users to take advantage of this potential in practice, since it can certainly make a significant impact on the ground. Within the current division of labour, which seems unlikely to change in the near future, it is usually design professionals who initiate proposals for physical form; and given the importance of initiative, which we explored in Part I, design professionals who are motivated to work towards more user-positive settings could in principle make a significant mark. It is therefore vital that we identify any common ground which users might share not only with the environmental lobby, but also with design culture too. Designers themselves certainly claim that common ground exists. The key ideological themes of mainstream design culture, after all, claim that designers' work supports users' aspirations in terms of community, safety, nature, health and art, as well as offering them enhanced levels of choice in their lives. Given our arguments for taking professionals' ideas seriously, we must explore these claims carefully, to see what potential for common ground between users and designers they reveal. That is our purpose in the next three chapters.

We are faced here with a complex task, and the first step in addressing it must be to develop a conceptual framework for addressing it in a clearly structured way. That is the subject of Chapter 7: 'Concepts for prospecting common ground'. Armed with this framework, we shall be able to analyse the extent to which, in practice, recent transformations live up to mainstream design culture's ideological claims, to see how valid they are in users' own terms. Are they more than convenient buzzwords to support the status quo? That is the question we shall explore in Chapter 8, 'Beyond buzzwords', and Chapter 9, 'Horizons of choice'. By the end of Part III, then, we shall have identified what common ground there is between the aspirations of users, designers and the environmental lobby; and we shall understand how recent transformations stack up in these terms. Armed with this understanding, we shall be poised to take well-informed action for change, to make good use of the new opportunities which the growing sense of ecological crisis opens up.

7 Concepts for prospecting common ground

If they are to succeed in the face of local economic resistance, moving beyond good resolutions to get significant changes on the ground, current international attempts to address ecological issues will need mass political support. In most countries, however, this is so far conspicuous by its absence: there seems to be relatively little current support for urgent change towards more sustainable urban forms, particularly in the rich countries which would need to change the most. In Britain, for example, attitude surveys suggest rather strongly that very few people allow their increasing awareness of global environmental problems to affect the practical organisation of their everyday lives.[1]

In part at least, this is due to the way the sustainability lobby presents its case. At the moment, as Andrew Ross points out, much of the sustainability debate is carried on in an ascetic and puritanical mode whose self-denying implications are deeply unappealing to many. This is emphatically not the way to build a mass support: 'The ecologically impaired need to be persuaded that ecology can be sexy, and not self-denying', as Ross himself puts it.[2] What we need, it seems, is an alliance between those who seek better-loved places and those concerned with ecological issues – an alliance responding to Ross's call for 'the hedonism that environmental politics so desperately needs for it to be populist and libertarian',[3] and recognising that in practice we can only get more sustainable places on the ground to the extent that they are better-loved too.

This need for a sexy ecology creates the preconditions for building an alliance between the environmental lobby and a wide range of everyday users. As we already saw, however, users and politicians also need to find as many allies as possible within the design professions, to take practical spatial advantage of the new potentials as quickly as possible. We therefore have to find transformations lovable enough to attract the active support of the widest spectrum of users, in ways which will also carry with them the environmental lobby and as many designers as possible. There are, however, all sorts of problems inherent in this necessary search for common ground, because the multiplicity of people who use given settings in building their everyday lives do so from a multiplicity of perceptions and interests. The anthropologist Danny Miller, visiting a London social housing estate, invites us to consider the sheer richness of the ongoing life-projects to be found behind its anonymous doors:

> An extraordinary feature of modern British life is the number and diversity of interests held behind these doors: a fanatic supporter of a football club lives next to a family that keeps an exotic range of pets; a follower of a pop group lives next to a political radical. Any one of these different cultural foci may become a central point of concern and identity. Television and the media continue to uncover the world champion hairdresser and the fancy-pigeon breeder, the bibliophile and the expert on coalholes.[4]

'An inescapable conclusion from such observation', he goes on to say, 'is that the culture of most people is of a very particular kind.'[5]

Cultural variations such as these may well lead to all sorts of different evaluations of any or all of the transformations we explored in Part II. A practical example may help to give a feel for the possibilities of disagreement here. In Figure 7.1, which shows a street in Goa, India, we see a house with a blank wall to the street, next to another whose verandah makes for a far more extrovert street edge. This difference is not simply a matter of aesthetic style. The blank wall shelters a Hindu family, whilst the verandah belongs to a Christian household; and the different interfaces here reflect and reinforce two different sets of cultural rules about the proper relationship between the domestic and the public realms.[6] Now, the members of these two families, with their different rules, might react

Figure 7.1 A street in Goa, with Hindu (left) and Christian (right) houses.

quite differently to a shift towards more 'passive' interfaces, of the kind we explored in Part II.

Faced with all this concrete, particular complexity, what chance is there of developing any shared conception of the qualities which better-loved places ought to have? How can we expect to find graspable patterns amongst the pop fanatics, pigeon breeders and coalhole fanciers of the real world? Do we not run a serious risk of losing the whole point of people's place-experiences if we try to understand them through some abstract intellectual framework drawn from outside sources? Does not the blanket notion of 'the user' conceal many important differences between all sorts of different users? Does not the generalised idea of 'given setting' beg important questions such as 'given for *whom?*' Clearly there are dangers here, of assimilating the positions of all users, in regard to all given settings, into some homogenised, sanitised commonality of experience, thereby losing the crucial capacity to make a critical analysis of who wins and who loses from today's process of urban transformation.

Discussion of such dangers formed an important area of debate in theorising such concepts as 'women' and 'race' during the 1980s.[7] During the 1990s, however, many of those who first saw the dangers, for example, of blanketing women together into a homogenised 'sisterhood' have come to acknowledge the other side of this particular coin: the political ineffectiveness which all too easily follows from moving towards a more closely-focused approach. As Mary Kennedy and Brec'hed Piette see it,

> The problem is, can we maintain the ideals of being truly sisters under the skin, allowing for the diversity of experience and culture arising from age, class, disability, ethnicity, race and sexual orientation, without splitting into fragmented, quarrelling groups while we lack the 'tangible politics' of a broad-based women's movement to link into?[8]

This caveat is also important in thinking about how people use given settings, if users are ever to influence significant change with the very limited power they have as individuals. From this practical perspective, it seems to me that we have to develop a frame of reference, for understanding users' interactions with given settings, in which a certain degree of richness is sacrificed in the interest of political empowerment. We should certainly not, however, brush the issues raised by cultural differences under the conceptual carpet. What we need is a frame of reference which is built on common experiential ground, but which is used in ways which are consciously held open to challenge when used in particular local situations. We shall come back to consider an appropriately open way of working in Part IV. For now, however, let us get on with exploring what common ground there might be between the mainstream culture of design, widespread users' desires and the specific concerns of the environmental lobby.

To search effectively for this common ground, we must first understand the nature of the relationship which all these various actors have with the given settings of their lives. In mainstream design culture itself, this relationship is conceptualised as a passive one: the 'consumption' of built form is seen merely as what happens to it after it is produced. Both from our structurationist perspective and from our everyday practical experience, however, we all know very well that users have an active relationship with the given settings of their lives. We constantly call on our settings to provide us with assistance in our everyday affairs, using them creatively as aids in doing the things we want to do, and even physically rearranging the bits over which we have sufficient control, to serve our purposes better. Everyone does this all the time, as Danny Miller points out when he comments further on

> the very active, fluid and diverse strategies by means of which people transform resources both purchased and allocated by the council into expressive environments, daily routines and often cosmological ideas: that is, ideas about order, morality and family, and their relationships with the wider society.[9]

The process of using given settings, then, is an active and productive one, to the extent that Michel de Certeau suggests that it could usefully itself be thought of as a form of production – an 'entirely different kind of production, called consumption'.[10] What is produced here, however, is very different from the built-form commodities whose production we have so far explored. The 'expressive environments, daily routines and cosmological ideas' which users create from and around their given settings are not externalised and discrete 'things' – they are users' own life-projects.

Both social and spatial factors are involved in constructing these projects. At the social level, factors internal to the user's subjectivity interact in this life-construction process with factors of an external nature. In terms of internal factors, the construction of any life-project is driven by desire, focused through the medium of ideology into a complex web of aspirations and internal resources such as knowledge and moral commitments. These internal factors govern the way in which available external resources such as money, cultural capital or frameworks of social support are deployed. At the spatial level, the resources offered by given settings themselves can be thought of as 'productive apparatus of a giant scale', in Henri Lefebvre's memorable phrase,[11] offering support and imposing constraints, at both physical and symbolic levels, on many aspects of the life-production process.

Users evaluate the changing patterns of opportunities and constraints which follow from urban transformations, in terms of experiential qualities related to their own webs of desire and aspiration. Drawing on ideas from ecological psychology,[12] the geographer Nigel Thrift offers a creative insight into the nature of these experiential qualities when he compares them to concepts like 'edibility'.[13] On the one hand, edibility is a property of the substance which is eaten, affected by physical factors such as its chemical composition. On the other hand, it is also a social matter, affected by cultural rules which define what counts as food, what tastes good and so forth, and also by the availability of resources such as recipes and cooking equipment. Though the nutritional value of witchety grubs is no doubt the same for indigenous Australians as it is for myself, for example, it is cultural factors which make them into food for them but repulsively inedible for me. The quality of edibility itself is therefore a quality of the relationship between the physical characteristics of the substance and the cultural rules and resources in whose context it is to be consumed.

From everything we have seen so far, it is clear that any user's evaluation of a given urban transformation stems from the interplay between a complex set of social and spatial factors. Evaluations depend not only on experiential qualities related to users' own patterns of desire and aspiration, but also on any new patterning of relevant opportunities and constraints which the transformation brings about, and on the array of other resources, internal and external, which the user can deploy to take advantage of the opportunities and to overcome the constraints. To seek new transformations, positive to a wide spectrum of users as well as to designers and to members of the environmental movement, we have therefore to investigate which patterns of desire might be shared across all these groups, and which patterns of resources might be available to them all. Let us begin by exploring the crucial patterns of desire which give each life-project its overall shape and direction. What grounds of desire can we find in common between a wide spectrum of users, designers and members of the environmental lobby?

At first sight, it might seem an impossible task to find much common ground at all. After all, we have just spent an entire chapter exploring the ways in which the current mainstream culture of design supports the high levels of resource consumption which exacerbate ecological degradation, and there is also a great deal of evidence to show that users and designers often have marked disagreements about the quality of recent transformations.[14] Because of this disagreement, and because design culture supports the requirements of the capital accumulation process so neatly, it is tempting to dismiss mainstream design ideology as 'mere propaganda'. This, however, would be a simplistic view, contradicting the arguments I have already made for giving serious attention to professional problematics. To see that common ground between users and designers must in fact exist, at least to some extent, we have only to recall what we already know about the way in which design ideologies are formed. As we argued in the last chapter, mainstream design ideologies are constructed to fulfil the need to make sense of the conflicts which inevitably from time to time arise between the developer-positive transformations in which they have to collude if they are not to go out of business, and their experience, as users, of the impacts these transformations have on their own everyday lives. Unless designers' own off-duty life-projects are completely different in kind from those of other users – and it is hard to believe that the counter-community of designers is *quite* that all-encompassing – then we might reasonably expect that there is at least some common ground to be found, and that professional ideologies might suggest where to look for it.

Because mainstream design ideologies form a conceptual bridge between the desires and aspirations of the 'professional' and the 'user' who co-exist in all designers' multiple subjectivities, the claims they make must relate to issues which matter in users' terms. The key ideological themes of choice, community, safety, nature, health and art, which we explored in the last chapter, must all therefore be linked to matters which are important in many users' lives. These issues, however, are expressed in distorted and mystified ways, so we certainly cannot expect to learn much about the common ground we seek if we simply take these ideologies at face value. These key ideological themes do not directly indicate the desires which are shared, but they do suggest rather strongly where we might look to find out which these are. Following their lead, we can begin to explore the senses in which these issues matter to users faced with recent transformations. To make this work, we have to step back from the ideological themes them-selves, and call on social theory once again to help us unravel what they can be made to reveal about the aspirations which drive their construction, and the experiential qualities which matter in these terms.

As we have seen, aspirations are structured through a process of making sense of one's life. Current widespread aspirations, therefore, have to be able to help users make sense of everyday life in today's capitalist context: a context 'given' by producers' attempts to keep the capital accumulation cycle working in a profitable way. Now the profitable operation of the capital accumulation cycle, as we already saw, is constantly threatened by problems of under-consumption. The producers of all sorts of commodities are under constant pressure to do whatever they can to promote ever-increasing levels of commodity consumption, in the interests of their own market survival. This, in turn, generates pressures for producing and promoting particular kinds of commodities.

Different kinds of commodities have different potentials in terms of the levels of consumption which users can sustain: some have inbuilt consump-tion limitations, whilst others do not. The very nature of food, for example, imposes obvious consumption limitations: there is a fairly strict, physio-logical limit to the number of steaks or chocolate bars which anyone can eat in a given time. To a lesser extent, the same is true of any commodity which is bought primarily for utilitarian reasons. For example, there is a limit to the number of tools I need for working in my garden, as long as I consider working in my garden purely as a practical matter of vegetable production. The moment I begin to think of my garden – or my clothes, or my car or whatever – as more than a strictly utilitarian affair, however, I start buying commodities at least partly for what Jean Baudrillard calls their sign value[15]

– to present a particular image of myself to others, and indeed to myself. In other words, I am buying them in part as ideological material for constructing subjectivity.

At this point, the situation changes in terms of consumption limits. Unlike the consumption of food, say, the desire to consume signs as an aspect of subjectivity-construction will only stop if I reach a state of perfect and permanent satisfaction with myself. Even in a monastery this is rare enough, no doubt. Elsewhere, given the complexity of the situations in which all users find themselves in modern everyday life, such a stasis seems unlikely ever to be achieved. This state of affairs offers producers fertile ground for developing strategies to promote consumption, by encouraging people constantly to question the validity of their own current subjectivities, through lifestyle marketing, fashion cycles and the like. Understanding the key role which this process of questioning plays in modern everyday life, we can begin to grasp the reasons why the ideological themes of choice and aesthetics have become so important today.

In this situation of constant self-questioning, for example, the act of choosing from amongst the sign-commodities on offer in the marketplace becomes central to the process of subjectivity-construction itself, so that choice can no longer be understood merely as a process of adjudication between the merits of alternative commodities. It has now taken on a crucial importance in its own right, to the extent that the ability of the field of commodities as a whole to offer choice begins to overshadow the qualities of the commodities in themselves. Indeed, even when goods are accessed through social mechanisms other than the marketplace – for example through Welfare State institutions – the same value of choice is seen as central to their quality. Bauman gives a nice example of what this implies:

> On my recent visit to Sweden I was told by quite a few ... intellectuals that – supremely efficient as it prides itself on being – the bureaucracy of the social-democratic state becomes ever more difficult to live with; and this is due to the limits it puts on individual choice. I asked my conversationalists whether, given choice, they would abandon the doctor currently assigned by the National Health, or seek another school for their children. No, was the answer: the doctor is excellent, and so is the school our children attend; why on earth should we go elsewhere? But, they told me in the next sentence, I missed the point. Quite obviously, the point was not the quality of doctor or school, but the feeling of self-assertion, expressed in the act of consumer choice.[16]

In the light of all this, we can see the key role which choice appears to play in structuring the open, exploratory aspect of desire in our pantheon of themes, and we can understand that it is therefore likely to be a major factor in the way in which users evaluate recent transformations. Unsurprisingly, therefore, diverse thinkers from various realms of social theory agree on pointing to choice as a widespread value today. Bauman, for example, tells us that 'choice has turned into a value in its own right; the supreme value, to be sure';[17] whilst Fred Inglis, focusing on the consumer from a cultural studies perspective, suggests that 'all happiness and fulfilment for this figure are defined in terms of *choice*, the key moral act which fills the consumer with purpose and identity.'[18]

In terms of the exploratory, creative aspect of desire, it is not a choice of 'things' which is desired, so much as a choice of futures for one's life-project – a choice of 'life-chances' as Ralf Dahrendorf puts it.[19] The way in which users construe the impact of particular transformations on the openness of their life-projects' futures is therefore likely to have a crucial impact on how the transformations themselves are evaluated. Those which, on balance, seem to offer more open futures are likely to be evaluated more positively than those which appear to close the future down. Even where, like Bauman's Swedes, particular users do not in practice want to change direction, they do want to know that they could do so if they wished. The importance of choice therefore has direct implications for the experiential qualities according to which urban transformations are widely judged. First, urban settings will be judged according to the variety of resources which they make available for users to choose from. Resources cannot be called on in the life-construction process, however, unless users can get into contact with them – either by 'going out' to them, or by somehow 'bringing them in'. The accessibility of the system of resources is therefore of crucial importance. Finally, users cannot effectively access resources of any kind, except by happy accident, unless they know how to find them. Legibility is therefore a third key quality against which urban transformations are likely to be judged.

The logic of our argument so far suggests that the three qualities of variety, accessibility and legibility are likely to be valued in common by many users and by many designers too; they point towards the common ground we seek. We must not, however, let our necessary enthusiasm for finding common ground blind us to the fact that there are also complex differences between different users' webs of desire. If we are to create new places which appeal as widely as possible across this spectrum of unpredictable and ever-changing differences, we have to find ways of making settings which have the potential to accept a wide and unpredictable range of alternative activities which different people with different values might want to pursue, and which also, at the symbolic level, actively suggest that these potentials exist. I shall call this quality robustness.

I am sure that the qualities of variety, accessibility, legibility and robustness do not fully capture all the possible interactions between the desire for choice and the pattern of resources which given settings make available. Nonetheless, this set forms a useful preliminary tool for analysing recent transformations in terms which are likely to be significant to a wide range of users.

In addition to promoting the importance of choice, the prominent role which the consumption of sign values plays in today's process of subjectivity-construction also gives a particular importance to the aesthetic dimension in all aspects of the life-construction process, for one important dimension of aesthetic experience – the cognitive dimension – is directly concerned with sign values – with the aesthetic aspect of the meanings which given settings have for their users. The cognitive level of aesthetic experience has therefore come to take on a particular and pervasive importance to many users, surrounded as they are by the intensive marketing of sign values which has become so central to the gearing-up of consumption. It is in the world of aesthetics, after all, that sign values are considered independently from values of other kinds. Or, to put this another way, 'aesthetics' nowadays means what we get when we make that separation. The particular emphasis on sign values in the consumption process is therefore expressed through a high level of concern for any given setting's cognitive aesthetic dimension – what Mike Featherstone rather memorably calls 'the aestheticisation of everyday life'.[20] This has a number of important implications for the cognitive aesthetic evaluations which users are likely to make of recent transformations.

First, it means that the experiential qualities through which people make their evaluations of given settings, in terms of all the other qualities we have explored so far, will probably themselves have important aesthetic components. Second, the importance ascribed to the cognitive aesthetic dimension also implies that judgements – good or bad – which are made according to other values will probably also be expressed in aesthetic terms. Types of transformations which are consistently experienced in negative ways for reasons quite outside the aesthetic realm, for example, are likely to influence the cognitive dimension of aesthetic values, so that the given settings which result from these transformations become widely seen also as aesthetically negative – ugly, crude or whatever.

Clearly we must pay careful attention to this two-way cognitive traffic between aesthetic concerns and the other aspects of users' agendas, but we must also be aware of the importance of another, non-cognitive or perceptual dimension within aesthetic experience as a whole. This perceptual dimension is concerned with the direct transactions which take place between the physical flux of the world and the physiological apparatus through which certain aspects of this flux are processed in perception. Because of its physiological basis, this perceptual level of aesthetic experience is largely cross-cultural in its relevance; this is a level at which aesthetic experience can be shared as common ground by people from different cultures, across large swathes of time and space. The quality of aesthetic experience at this perceptual level depends on the interaction between the physiological apparatus of perception and the patterning of the vibrations of light, sound and so forth which make up any given setting's physical flux. The physiological apparatus itself appears to be set up in such a way that physical flux which is patterned above a certain threshold of complexity is perceived negatively as chaotic, whilst patterning below a second threshold is also perceived negatively, as boring. Between these thresholds lies a complexity-range which is perceived as positive richness – a further important quality of urban settings, for users and designers alike.

The importance of the 'aestheticisation of everyday life' means that aesthetic experience is not merely a kind of icing on the cake of urban experience as a whole. Rather, it unavoidably permeates users' agendas throughout. This means that aesthetics is sexy; if we want to build common cultural ground which is solid enough to take strong foundations for better-loved futures, we have to engage fully and enthusiastically with this aesthetic dimension. From this perspective, therefore, we can only deplore the reductionist view that aesthetic qualities are merely by-products of other more important issues, which we find in much otherwise interesting work in fields such as community development and community architecture. This anti-aesthetic bias does not help to build the common cultural ground we seek.

To summarise, then, we can see that the ideological themes of 'choice' and 'aesthetics', as we expected, are not simply dreamed up by designers as propaganda to support the transformations in which they collude. We can now see that these themes are likely to have a central importance for many everyday users, and are therefore likely to affect the ways in which other widespread levels of desire are developed. In particular, the importance accorded to choice as a value in its own right, independent from the things or states which are chosen, creates paradoxical difficulties for the choosing process itself. In situations where choice is the supreme value, people want to exercise choice in as many aspects of their lives as possible; including, necessarily, choosing the criteria by which their choices are to be made. This potentially leads to an infinite regress of choice – a choice of ways of choosing how to choose, and so on *ad infinitum*, offering no obvious basis for making sense of any particular choice. And yet sense has somehow to be made – in practical terms, the chooser has to find some way in which choices can be validated, whilst forgoing as little choice as possible in the process.

Different individuals, of course, find all sorts of ways of squaring this potential circle, but two approaches present themselves as particularly salient, and help us to understand the sense in which the ideological themes of health and community, which are so important in the culture of design, are also of widespread importance in users' terms. The first approach is to look inwards at one's own body; grounding one's everyday choices in a desire to maintain one's health, and therefore one's capacity to carry on as an effective chooser. The other approach is to look outwards, in search of a community of other people who would agree the kinds of choices one makes, and by agreeing validate them. Viewed in this light, the themes of health and community come into a new focus.

The state of a person's health depends both on the qualities of the basic environmental media of air, water and soil on which all life depends, and also on the 'operating programme' in force for the body itself. Urban transformations clearly impact on both these factors. In health terms, what matters about the basic environmental media is their freedom from noxious pollution – what is usually termed their cleanliness. This is also important at a symbolic level, where 'nature' is nowadays often taken as an important symbol of purity and cleanliness; hence the important links between the ideological themes of health and nature which we noted in the last chapter. The evaluation of recent transformations in health terms, then, is often affected by what Farmer calls a 'green sensibility',[21] by which places are valued to the extent that they look or feel 'green' or 'natural', because of the presence of planting and urban wildlife. The quality which matters here, then, might best be called 'biotic support'.

The patterning of urban space also affects, somewhat less directly, the effectiveness of the regime through which people maintain their own bodily health. What matters here is the degree to which the structure of urban space encourages the bodily exercise which comes from active use. The major requirement, in these terms, is for given settings which can yield high levels of the key qualities of variety, accessibility, safety and robustness which people seek in the course of everyday life, with the minimum need for outside aids such as mechanical transport, so that the body can pleasurably be called on as a key life-construction resource.

The grounding of choice in a concern for the user's own bodily health, then, places still further emphasis on the importance of the qualities of variety, accessibility, safety and robustness in urban settings, and also draws our attention to the practical importance of the cleanliness of the basic environmental media of air, soil and water, and to the aesthetic importance of biotic support. Which further further implications might follow from adopting the alternative strategy: grounding the choosing process in some wider sense of community support for the choices we make?

In this second, community-orientated approach, it is the sense of belonging to some community larger than oneself which enables one to validate one's choices against the approval of other members of the community concerned. In situations where choice is the supreme value, however, any sense of community has to be constructed in ways which differ markedly from the traditional 'communities of fate' into which, without choice, one is born, and within which one's choices are strongly constrained by rules which are presented as given, and over which one has no direct control. In exploring the differences here it is helpful, I think, to distinguish between two complementary aspects of the overall notion of community. On the one hand, there is the aspect which is developed through concrete, face-to-face interactions. On the other hand, in almost all real situations, there is an aspect of community whose construction takes place in symbolic or spiritual realms – what the sociologist Benedict Anderson calls 'imagined community'.[22]

So far as face-to-face communities are concerned, where choice is the supreme value, it must extend to the ways in which communities themselves are constructed. This fosters the development of communities which are voluntarily built up by their members, each seeking others to whom they attach themselves for mutual support. The 'community of choice' becomes more important, at least subjectively speaking, than the 'community of fate' for many people; and in so far as given settings might affect the process through which imagined communities are formed, those which make it easier are likely to be evaluated more positively than those which make it more difficult. Even in the context of freely-formed communities of choice, however, community approval and individual choice are always potentially in conflict. What is sought is a balance between them. The exact nature of the balance is itself a matter of choice, but in a situation where choice is the supreme value there must be a strong tendency towards seeking community approval in ways which, overall, reduce choice as little as possible. Communities in which constraints on choice are reduced to zero, so that the individual's own choices are always approved, can only be constructed in the mind, as communities in which only the imagined dimension is present. Zygmunt Bauman describes these imagined communities as follows:

> belief in their presence is their only brick and mortar, and imputation of importance their only source of authority. An imagined community acquires the right to approve or disapprove in the *consequence* of the decision of the approval-seeking individual to invest it with the arbitrating power and to agree to be bound by the arbitration (though, of course, the reverse order must be *believed* to be the case to make the whole thing work).[23]

Even the most traditional 'communities of fate', of course, have important 'imagined' components, but the point about the 'imagined community' as Bauman discusses it, is that 'belief in their presence is their *only* brick and mortar,' as he puts it. They are constituted only at an ideological level, and therefore they can only be constructed through symbolic resources.

Because everything has a symbolic dimension, anything can in principle be called on in the community-construction process. Urban settings, however, provide a particularly rich and potent source of material here: they are full of other people, they are large and unavoidably noticeable, they affect our daily lives in pervasive ways, and because they physically envelop individuals together, they can easily be seen as spatial analogues of communities themselves.

The presence of other people in urban settings is a key resource here; not only for the physical meetings on which face-to-face community depends, but as symbolic resources for the construction of imagined community too. The importance of this more 'detached' symbolic relationship is central to understanding why 'people-watching' is an important aspect of city life to many citizens. It helps to explain, for example, why William H. Whyte, observing performances in New York's First National Bank Plaza, found himself looking at 'people looking at people who are looking at people who are looking at the show',[24] and why, in the very different setting of the Amsterdam pavement café, Jan Oosterman found the same:

> The most important and favourite activities of people on the sidewalk-café are watching people go by, being entertained by street life and inhaling the atmosphere of the city. Chairs are always placed towards the street, like chairs in a theatre are placed towards the stage.[25]

The only requirement for enjoying these pleasures is a supply of other people, who merely have to *be there*: 'Generally speaking, one does not want to get involved with anybody passing by. People mainly come to see the spectacle in front of their eyes.'[26] The vitality of public space, measured according to the level of pedestrian encounters which it offers, is clearly the most important potential resource in these terms.

In addition to the physical presence of other people, the spatial characteristics of given settings also offer rich symbolic resources for users to call on in constructing the imagined community they long for, to make sense of their everyday lives. We are concerned here with the degree to which the characteristics of these morphological elements lend themselves to interpretation, by those who long for imagined community, as being appropriate to the imagined community they desire. 'Is this appropriate as *our* place?' is the important question here, whatever different implications 'our' might have in specific instances. All things being equal, urban transformations which increase the likelihood of an affirmative answer are likely to be evaluated more positively than those which reduce it. Places which seem positive in these terms are often said to have an appropriate *identity*.

There is a problem here, however: in the light of recent and current events in many parts of the world, we have to be very aware of the dangers which can all too easily arise from fostering an 'us' versus 'them' sense of imagined community, demonising 'the others' in today's multi-cultural societies. Appropriate identity, like cleanliness and biotic support, helps users to validate the choices they make in their everyday lives; but the imagined community which it supports is imagined as conflict-free – it is constructed to support one's own choices, after all, rather than to dispute them. This means that people who live their lives with a major focus on imagined communities, rather than on face-to-face interactions with other people, develop to that extent fewer skills in making compromises, coping with ambiguities and so forth. In practice, this all too often breeds a stereotyped, intolerant view of the 'them' outside one's own imagined world. All 'they' do, in these terms, is place constraints on the choices 'we' can make. The upshot is that we feel little or no responsibility for their well-being, and we 'know' (from our viewpoint within our imagined 'us') that they feel about us in the same uncaring way. Mutually, therefore, we are all too likely to develop conceptions of each other as threats to be feared.

This dynamic helps us to understand the current importance of the ideological theme of safety, for it seems to have led to a situation in which concerns about safety have shifted from a primary focus on 'the accident' towards 'the crime' as the prime source of danger. The 1994 British Crime Survey,[27] for example, makes it amply clear that even in the UK – far from the most crime-ridden place in today's world – more people are 'very worried' about becoming victims of crime than about losing their job, illness in their family or being injured in a road accident. In this situation, many people have developed ambiguous perceptions of the ways in which their own safety is affected by the presence of others. A heightened concern about 'them' as potential threats co-exists uneasily with an awareness of 'us' as potential helpers in times of danger, or as potential deterrents to ill-intentioned acts. It is the former view, rather than the latter, which creates widespread opportunities for capital accumulation; through the production and sale of a broad range of goods and services, from locks and chains to companies which specialise in cleaning up the scenes of violent crimes. Not surprisingly, therefore, the resources of the advertising media have been deployed to gear up demand, by exacerbating people's fears in this area – an extremely effective campaign which in Britain, for example, had already supported the growth of the 'security sector' to a £2 billion-a-year industry by 1995.

There is no doubt, then, that safety is indeed a key – even an exaggerated – concern for many people in modern urban life; and that perceptions of safety in urban settings are closely linked with the presence or absence of other people there. Whether one sees others as threats or as potential allies, the presence of other people matters. The extent to which a given setting is 'peopled' – its vitality – is therefore an important quality so far as many users are concerned, whether they see this in positive or in negative terms.

By this point, we can see that the strategy of following up leads from the mainstream ideologies of design has indeed helped us, as we hoped, to identify some of the common ground we seek. To summarise, we have identified nine key qualities – variety, accessibility, legibility, robustness, richness, identity, vitality, cleanliness and biotic support – which probably matter both to many users and to many designers. We can also begin to see the basis for a certain number of links with the values of the environmental lobby. First, there are obvious connections through the quality of cleanliness, in regard to the basic environmental media of air, soil and water. Second, the quality of robustness has positive implications for energy conservation, since high levels of robustness are associated with places which do not have to be wastefully demolished and reconstructed every time human purposes change. Third, the aesthetic appreciation of greenery and urban wildlife, though doubtless superficial in ecological terms, still offers a cultural basis to build on. Even on the most optimistic interpretation, however, this does not yet look like an adequate foundation for building a mass movement for a sexy ecology – as we have seen, there is as yet little evidence of any widespread willingness to change patterns of consumption in order to behave in ways which are more ecologically benign, by reducing the polluting consumption of scarce resources: the 'supreme value' of choice, expressed through consumer choice, wins out over ecological values every time. How, then, might more common ground be created here?

The potential for promoting new transformations towards a sexy ecology – transformations which seem lovable to members of the environmental lobby as well as to designers and users – stems from the fact that the qualities which any user ascribes to the experience of a given setting do not usually derive from some direct interplay between the user's pattern of desire and the physical characteristics of the setting itself. As we already saw, the quality of this experience depends also on the other internal and external resources which users bring to bear, to wrest the best qualities they can from the physical characteristic of any given place. If we are to build new futures founded on the common ground between users', designers' and environmental values, therefore, we have to seek transformations which can offer high levels of the key qualities we have explored, with a minimum need to consume scarce resources and a minimum creation of pollution in the process.

Such transformations would also have the potential for support from a still wider constituency of users, because they would offer better qualities, in terms which are widely valued, to people who have the most limited resources: the poor, the elderly, children, and all sorts of other disadvantaged groups who, taken together, probably constitute a majority in most societies today. And even for the richest users, places which offer the best qualities for the least resource expenditure must necessarily free up resources for other aspects of the life-construction process.

If such transformations could be brought about in reality, then, they would offer a hedonistic route towards a more ecologically benign future. But what does all this mean in concrete terms? What would such transformations be like in practice? We know that we cannot start on this road with a blank sheet, but have rather to proceed through the transformation of existing mainstream types. But what should be changed and what conserved in these new transformations we seek? If we are to make creative and responsible proposals, we have to know how recent transformations have affected the amounts and types of resources which users have to deploy to avoid losing out in terms of the key qualities we have explored. In the next two chapters, we shall evaluate recent transformations in these terms.

8   Beyond buzzwords

In the last chapter, I argued that choice is nowadays 'the supreme quality' for many people, 'irrespective of the things or states which are chosen' as Zygmunt Bauman puts it.[1] But this does not mean that these 'things or states' do not matter at all: it is clear that choice will be exercised not only for its own sake, through the open, exploratory aspect of desire, but also in attempts to satisfy the more focused aspirations for positive experiences of community, safety, health, biotic support and aesthetic richness. In a sense, we can see these more focused aspirations as forming prerequisites for meaningful choice. How do recent transformations help or hinder in these areas?

When we consider the supports which given settings offer for constructing experiences of community, we have to take account of two complementary aspects of the overall concept of community itself. On the one hand, we have to consider how recent transformations have altered the capacity of given settings to offer resources for building community face-to-face. On the other hand, we must also consider how supports for constructing the 'imagined' dimension have changed. In contrast to the imagined dimension, the face-to-face aspect of community can only be constructed through close, extended contacts. In making these contacts, people call on social rules: all cultures have rules which regulate the conduct of encounters, and urban life would be intolerable for most people without some version of what Erving Goffman calls 'civil inattention',[2] which enables us politely to distance ourselves from others in most situations. These cultural rules prescribe the situations in which close, face-to-face interactions are to be sought – interactions with 'neighbours', or in shops or various kinds of communal settings from religious to leisure, for example. Many recent transformations have made such contacts more difficult than they used to be, particularly for users with few external resources.

At the largest scale, for example, we have already seen how the dispersion of overall settlement patterns, together with the reduction in levels of pedestrian flow throughout much of the public space network, have together led to greater distances between these kinds of interaction-settings. The reduction in the potential for community-building contact here is exacerbated by transformations at the smaller scale of the interface between buildings and public space. Here the shift from 'active' to 'passive' forms reduces the chances of contact with neighbours, for example; as an Angell Town resident laments:

> When I lived in a bedsit, it was horrible in lots of ways, but at least I could look out of the window and see people coming and going. If I knew someone, I could wave and maybe start up a chat. Here I can't see out at all, except at my own garden.[3]

The extent to which this reduction in the chances for culturally-appropriate face-to-face contact matters in practice, to particular users, seems largely to depend on the external resources which they have available for compensatory action. The importance this has for particular users must, of course, depend on the level of external resources which they can marshal to aid their personal mobility. Those with easy access to public transport or to a private car whenever they want it, for example, must be less affected by the shift towards enclave structures than those without. The experience of the writer Jonathan Raban demonstrates the importance of mobility in this regard:

> I live in a community whose members are scattered piecemeal around London (some of them live outside the city altogether); the telephone wire is our primary communication, backed up by the tube line, the bus route, the private car and a number of restaurants, pubs and clubs. My 'quarter' is a network of communication lines with intermittent assembly points; and it cannot be located on a map.[4]

This roughly approximates to my own situation and, no doubt, to that of so many other likely readers of these pages that it might be in danger of being taken as a current universal, so that any other view comes to feel like sentimental nostalgia. It is very important to realise, however, that this kind of community is simply not practicable for users with more limited mobility. Elderly people, for example, are often in practice less mobile than their juniors; so it is not surprising that the 1993 American Housing Survey shows that people over 65 years old are almost four times as likely, compared with younger people, to choose a housing location because it is convenient for access to friends or relatives.[5] This effect is compounded in situations with even less welfare support than is available in the United States, and where poverty reduces the other resources available. The words of Doña Maria, an impoverished elderly Mexican woman resisting well-meant attempts to relocate her to 'better' conditions after an earthquake, may help to bring this home:

> No. Under no conditions would I accept to be sent somewhere else. Not even if they promised me paradise. Do you know why? Because I have

lived here since I was little, everyone knows me, they more or less know my situation: I live alone because my children are scattered in other places. When I get ill, the neighbours help me a lot. They bring me food, they tell me how to take my medicine and they take care of me because they know who I am: Doña Maria. Somewhere else, I am no one or even less . . .[6]

In terms of their impacts on the resources available for the construction of communities of choice at the face-to-face level, then, recent transformations seem negative at the largest scale of the overall settlement pattern. How have the smaller-scale transformations of the public space network, with the proliferation of spatial enclaves, affected this issue?

First, the removal from the enclaves themselves of through vehicular traffic probably does have the positive effect of reducing barriers to potential interaction between people who live there. A comparative study of three San Francisco streets, by Donald Appleyard and Mark Lintell, helps us to see the relationships here.[7] The streets, referred to as Heavy Street, Medium Street and Light Street because of their very different levels of vehicular traffic, were parallel and near to each other, and nearly identical in physical design. Heavy Street had 15,750 vehicles per day, Medium Street had 8,700 but Light Street had only 2,000. These differences had important effects on the pattern of contacts between the people who lived in the three streets.

> With regard to *neighbourliness*, the inhabitants of Light Street had three times as many local friends and twice as many acquaintances as those on Heavy Street. The network of social contacts . . . shows that contact across the street was much rarer on Heavy Street than on Light Street. The friendliness on Light Street and the loneliness on Heavy Street came out clearly, especially in the responses of the elderly.[8]

Even on Medium Street, the severance effects of the traffic were strongly felt: '"It used to be friendly", one interviewee told us; "what was outside has now withdrawn into the buildings. People are preoccupied with their own lives."'[9]

The construction of any imagined community of choice, however, depends partly on a supply of alternative models and images of other people, from which one can decide which are 'significant' to oneself, and which are not. Potentially, encounters with other people in highly-connected public space can provide a plentiful supply of such images, which people – particularly young people – can experience through such settings as pavement cafés. As Jan Oostermann found in Amsterdam, for example, 'most visitors of the café are accompanied by a friend and practically all of them enjoy discussing the appearance of other people who walk by.'[10] They also enjoy displaying their own images to others: 'the sidewalk café is used to show off one's popularity, beauty or personal happiness, for example by kissing in public.'[11]

With the shift from highly-connected grids of public space towards hierarchical systems, creating large numbers of enclaves, opportunities of this kind are diminished, because the variety of available images is likely to be very limited within any particular enclave. These images can, of course, be encountered through such media as TV or magazines, but here they are unavoidably orchestrated, pre-digested and selected by producers. High levels of choice require more openness than that, so the 'uncensored' quality of casual encounters in public space forms a key resource here, particularly for younger people who are likely to be the most experimental in this regard. The reduction in likely encounter rates in most areas of the public space network, brought about by the transformation towards enclaves, is clearly negative in these terms.

The scope for constructing imagined communities, conceptualised around the distinctive identities of particular places, whose potential importance we explored in Chapter 7, has also been markedly affected by recent transformations. At the level of the public space network, for example, the shift from grids towards hierarchies, with the proliferation of enclaves, has important implications. What we see here is a kind of spatial onomatopoeia, in which the capacity for the enclave to be seen as a positive 'figure' within the general urban scene, distinct from all around it, makes it easily seen as belonging to an 'us' distinct from a 'them' outside. The cultural rule which encourages many users to make an imagined connection between enclaves and 'community' is therefore as enticingly easy to learn as the one which makes the word 'bang', for example, stand for an explosion.

The move towards enclaves, then, is easily seen as offering positive support for the construction of imagined community, particularly as its exclusion of through-traffic probably does make for easier face-to-face contact between neighbours, and between their children too. The problem with this approach, however, is that it supports exactly that 'exclusive' view of imagined community – 'us' versus 'them' – whose social dangers we remarked in Chapter 7. Conversely, the reduction of public space itself from positive 'figure' to negative 'ground', in many users' perceptions, reduces any capacity which the forms of public space networks might have to support a more inclusive notion of us-ness. Overall, then, public space networks have been transformed in ways which have many negative aspects from the point of view of supporting non-divisive imagined community. But the public space network itself is not the only medium through which symbolic support might be offered in this regard. How have recent transformations affected other elements through which such support might be offered?

The transformation from 'local' building types towards a more 'global' vocabulary – manifested, for example, in the global proliferation of office towers, out-of-town malls and detached suburban houses – is clearly one factor which reduces the distinctiveness of any particular local identity. Further reductions are brought about through the widespread shift from locally produced building materials to those bought in a global market. Occasionally, paradoxically, the existence of this wider market makes it possible to employ 'traditional' materials which are no longer locally available; as when the local distinctiveness of roofing materials in Britain's rural Cumbria, enforced through local conservation policies, can only feasibly be maintained through the use of slates imported from Latin America. More often than not, however, economic pressures result in building materials selected primarily for their capacity to aid the capital accumulation process. In practice, there is no doubt that the overall result is to reduce the surface differences between places, reducing the uniqueness of any particular local identity still further.

In particular situations, where the past has negative connotations for local people, this shift away from previous forms and surfaces may be regarded in a positive light. This was certainly what the architect Marilys Neponsechic found out as the result of winning a competition to design affordable housing for Mount Olive, a turn-of-the-century, African-American neighbourhood in Delray Beach, Florida. The competition entry had been developed through the use of types drawn from the place's historical context, as the architect explains:

> We responded by turning to examples of the Shotgun, a house type brought to the Florida Gold Coast by its earlier African American (Caribbean) settlers – and the Charleston single house, a type indigenous to a region with history and climate greatly similar to those of Mount Olive . . . We believed the historic authenticity of the Shotgun and the urbanism and climate appropriateness of both types were vehicles for expressing the most compelling elements of Mount Olive . . . It was our hope that residents would find self affirmation and empowerment in our Proposal.[12]

The reaction of the prospective residents, however, was far from positive:

> Prospective African-American residents of Mount Olive have refused to commission any building with a resemblance to the quarters of their ancestors, explaining that houses with such clear lineage to a slave past could only serve to stigmatize and marginalize them further.[13]

In situations where the past has less negative connotations, however, the reduction in the distinctiveness of local identity which so often results from recent transformations is often deeply regretted, by lay users and design professionals alike. In a study which compared Aldo Rossi's Centro Direzzionale in Perugia, Italy with a housing project in Reading, UK, designed by my colleagues and I, Lucia Vasak asked the question, 'Do you feel it is important for such a development to support/reinforce the existing character of the city?' She found that 95 per cent of the UK respondents and 99 per cent of the Italians answered with a straight 'yes'.[14] In Britain, at least, there also seems to be widespread support for this view in professional circles. Research into the attitudes of a broad range of professionals, initiated by the Department of Environment in 1994, showed very clearly that the reduction of local identity is widely regarded as one of today's key urban design problems.[15]

Most of the transformations we have so far reviewed have had negative impacts, both on the face-to-face and the imagined aspects of community-building, particularly for those users with the least ability to bring other resources to bear. The shift from grids to enclaves, however, has effects which are likely to be seen by many users as positive in these terms: at a face-to-face level, the removal of vehicular through-traffic which it promotes has beneficial effects on the potential for neighbourly contacts, whilst many have also learned to associate the closed spatial structure of the enclave with 'community' at the imagined level. Overall, then, the impact of capitalist transformations on the supports which given settings can offer for users to construct communities of choice is a complex and contradictory one. What impacts have these transformations had on perceptions of safety, the second key issue in many users' minds?

Recent transformations from compact to dispersed settlement forms and from grids to hierarchies have been taken up as positive features, in safety terms, by those advocating the 'gated enclave' as a spatial type in recent years.[16] The principle here is that fences, gates and security guards are used to keep out all but those who have 'legitimate business'. No longer can anyone else explore the enclave concerned, or pass through it on their way to somewhere else; so anyone who might be seen as posing a potential threat to those inside is excluded. Any supposed safety advantages which this might bring for those inside the enclave, however, are bought at the expense of those left outside – necessarily the majority, given the relatively greater cost of constructing, managing and policing the gated settlements themselves – since the relatively influential and law-abiding citizens on the inside are no longer motivated or, probably, even able to make any contribution to the safety, or the sense of safety of conditions outside. There are further disadvantages, for even those inside buy their extra security, all things being equal, at the expense of a reduction in resources for constructing both face-to-face and imagined communities, which we have already discussed as flowing from the enclavisation process itself. Systems for the effective screening of outsiders cannot be afforded, in any case, by precisely those people, already disadvantaged in various ways, who are most affected by crime, and most fearful of its impacts on their lives. How does the enclave approach affect *them* from the safety point of view?

The 'keep people out' view suggests that the shift towards enclaves should have positive effects on safety and its perception. Certainly there is a body of research which at first sight appears to support this view. The work of Barry Poyner and Barry Webb,[17] for example, relating layout to reported burglaries, suggests that culs-de-sac are safer from burglary than through streets. The evidence here, however, seems open to reinterpretation. Ian Washbrook, for example, has brought together all Poyner's and Webb's crime incident plans to compare the distribution of crimes for the entire study area which they covered, showing whether these occurred in culs-de-sac or grid-planned areas.[18] Presented like this, the results suggest a higher incidence of crime in the cul-de-sac areas as compared with that in the grids; except for burglaries of cash and jewellery, which appear to be equally spread between the two.

The tentative support which this gives to the alternative 'invite people in' approach is reinforced by a great deal of work carried out by Bill Hillier and his colleagues at UCL. Analysing where burglaries took place in London's Marquess Road Estate in 1988, for example, Hillier discovered that this type of crime was most frequent in relation to properties in the most segregated parts of the area, and least in those located on the more integrated spaces.[19] These findings, as Hillier puts it,

> cast quite fundamental doubts on the whole concept of 'defensible space', at least insofar as one of the main assumptions behind it is that the elimination of natural movement and encounter within housing estates will increase safety. Advocates of defensible space from Newman onwards seem to believe that the criminals seeking victims are part of the passing crowd, and that strangers are therefore in principle dangerous. Something like the opposite appears to be the case. The natural presence of people may be the primary means by which space is policed naturally. The more you eliminate this, then the more you create danger once a potential criminal has appeared on the scene. It is true that people behave more 'territorially' in segregated spaces. The more segregated, the more likely one is to question the presence of strangers. But this is associated with feeling more unsafe. No one feels the need to question strangers passing down a street. On the contrary, their natural presence increases the sense of security.[20]

In my experience, this line of reasoning fits well with the practical experience of those most at risk from crime itself – the least well off, the elderly and members of ethnic minorities. Residents of the Angell Town estate in Brixton, South London, include disproportionate numbers of people in all these vulnerable categories, so we should expect fears about safety to be high there. As one part of a large survey, which included amongst its respondents 70 per cent of the residents of the block with the worst crime reputation, the following questions were asked:

Do you think the streets on the estate would be safer if they had more people walking in them?

There might be more pedestrians in the streets if we make it easier for non-residents to walk through the estate to get home. How would you rate this idea?[21]

Over 80 per cent of the respondents answered 'yes' to the first question and, perhaps more significantly, answers to the second showed that some two thirds of respondents thought that having more non-residents walking in the streets would make the place safer. This is particularly interesting given that Angell Town, at that time, had a very bad reputation in police circles, and that it was designed very much as an enclave, with little spatial integration with the areas around it. This perception amongst users that 'people equal safety' is not only confined to the users of disadvantaged areas, nor is it specifically British. A survey conducted by students at Canada's Carleton University, for example, showed that 'isolation or few people around' was one of the four most cited reasons for feeling unsafe on the university campus after dark.[22]

If the presence of other people, on balance, has a positive impact on both the perception and the substance of safety, then clearly the length of time during which other people are present must also be important from this point of view: situations where people are around all the time must be more positive than those with long periods of inactivity. The transformation from a fine to a coarse grain of local use is a crucial factor here, for mono-use zones tend to be emptied of occupants for part of each day and each week, while areas of fine-grained mixed use offer the potential for mutual surveillance between buildings which are occupied at different times.

The extent to which this potential is achieved in practice depends in part on the design of the interface between each building and the public realm. The way in which the nature of this interface affects users' feelings of safety has been explored through recent studies in Oxford. Here, groups of men and of women were each taken on accompanied walks through areas of the city with different layouts and different patterns of use. They were asked to comment on their perceptions of threat in different physical settings along the way, and the results were compiled into 'fear maps'.[23] As we might expect from the arguments above, areas with lots of overlooking from adjoining buildings were typically rated as far safer than those without. The best-surveilled areas, in these terms, certainly included culs-de-sac, a fact which at first sight gives support to the 'safety through enclaves' idea. But how does this work?

The arrangement of the cul-de-sac itself offers high levels of surveillance from buildings: surrounded by building fronts, and therefore overlooked from the buildings by people whose right to be there is itself legitimated by the fact that they have been allowed into the private spaces of the buildings themselves, it is easy to understand how anyone within the enclave itself is likely to feel safe. In contrast to grid-planned areas, however, this double-sided surveillance is not possible everywhere within any hierarchical area. With all the 'active' building fronts facing on to the enclaves, the other public spaces which link each enclave to the outside world are necessarily presented with private backs. In any society where privacy matters, this means that they are effectively lined with blank privacy-screens of one kind or another, with very few doors or windows to contribute signs of life. With much reduced surveillance, it should not surprise us that feelings of unease are more common here (Figure 8.1).

These uneasy feelings mean that every route from the 'home' enclave into the settlement as a whole involves venturing into less-safe-seeming areas. Even for people who feel least vulnerable, this acts as a further disincentive to using the city in an open and exploratory way, except through the medium of the private car which acts as a protective armour around a moving bubble of private space, used to traverse the relatively threatening realm of public space as quickly as possible.

The wariness with which users regard blank walls to public space seems to be justified in practice, in terms of the incidence of crimes associated with such situations, as revealed by studies which explore the particular routes by which burglars break in. Talking to burglars themselves, for example, Bennett and Wright found that the preferred approach was usually from the back.[24] This seems amply borne out by the crime statistics. The 1992 British Crime survey, for example, found that almost twice as many burglars

Figure 8.1 An unsurveilled space which many people found frightening.

got in through the back in comparison with the front,[25] whilst Hillier's work on the Marquess Road Estate also found that the burglar-vulnerability of particular properties depended on the level of 'static' surveillance enjoyed by the routes leading towards them: the greater the surveillance provided from dwellings fronting on to the route concerned, the lower the level of vulnerability. This re-emphasis on Jane Jacobs's original insights about the complementarity of 'strangers' and 'insiders' in creating safe places, which is supported by other studies,[26] throws into sharp relief the negative safety implications of the transformation from grids and perimeter blocks – a typology which generates streets with continuous surveillance from building fronts on both sides – towards layouts which face blank backs on to public space.

To summarise, then, the urban transformations of recent times, at all scales from the building/space interface through the grain of land use and the network of public space to the overall settlement pattern, seem mostly to have negative impacts both on safety and on its perception. The search for capital accumulation has effectively generated a built world in which the safety of small enclave islands has been maintained at the expense of increased perceptions of fear in the whole, affecting particularly those who have the least protection through external cultural rules or through internal resources of physical strength and the willingness to use it in violent ways. Issues of gender, ethnicity and age are all important here.

To make matters worse, this situation is inherently unstable; though the fear of crime is almost always far greater than the level of crime itself, Lea and Young suggest that it may itself act as a self-fulfilling prophecy, to increase the very level of crime which is feared.[27] The problem is that an increased fear of crime, particularly when it is sensationalised by the communications media, leads both to a situation in which fewer people venture into public space, and to a lower willingness to identify criminals, to testify in court and so forth. This in turn leads to a reduced perception of risk on the part of criminals themselves, and therefore to an increase in actual crime. Picked up again by the media, this still further increases the fear of crime; and so on and on in a vicious spiral of decline.

So far, then, our analysis of recent typological transformations from the points of view of community and safety has brought out a dismal picture of mismatch between urban form and users' agendas. But how do these transformations stand up to scrutiny from the health perspective?

Urban transformations have one of their most direct effects on people's health through their impact on the cleanliness of the basic environmental media of air, water and soil; in particular through air pollution. The climatologist Derek Elsom, for example, suggests that currently, worldwide, 'the health of 1.6 billion people may be at risk from poor urban air quality',[28] and the situation is worsening year by year. Working with World Bank predictions, Elsom tells us where we currently seem to be headed: 'If the growth in demand for transport and electricity were to be met with the techniques currently in use, emissions of the main pollutants derived from these sources would increase five-fold and eleven-fold, respectively, by about 2030.'[29] Though much of this increase is due to the growth of world population, the demand for transport and electricity are also directly affected by spatial transformations at a range of physical scales. The shift from compact to dispersed settlement patterns, the move from grids to hierarchies in the realm of the public space network and the transformation of building form from terraced masses towards free-standing pavilions are all important here.

The consumption of fossil fuels, for example, is geared up through the increasing dispersion of settlement patterns, which spaces urban facilities ever further apart. As this process develops, it becomes less and less feasible to achieve a given level of choice amongst the city's facilities without extensive use of motorised transport: low-emission feet and bicycles will no longer do. As dispersal increases, densities fall below the level at which it is financially feasible to provide public transport as a marketable service, without levels of subsidy which have so far proved difficult to achieve within a capitalist framework. The extensive use of private cars, for those who can afford them, is the only remaining option for maintaining the user's level of choice amongst the now-dispersed facilities, with implications we shall explore in greater depth in the next chapter.

This move away from feet, bicycles and public transport towards the private car leads to greatly increased levels of energy consumption. Figure 8.2 compares the energy used per passenger-kilometre for these different transport modes: we can see that private cars, even with a full complement of passengers, use about ten times as much as bicycles, for example, for a given journey length. If a four-seater car is occupied by only one person, of course, the comparison is much worse: now the car uses forty times as much as the cyclist. To make matters worse again, the car uses non-renewable fossil fuels in place of the renewable food resources which fuel the walker and the cyclist. If vehicle technology stays constant, then increasing dispersion leads directly to increasing fuel consumption; the relationship here is a very direct one (Figure 8.3). Increasing levels of fuel consumption are also triggered by the transformation of public space networks from grids to hierarchies. To the extent that it allows increased traffic speeds, this transformation makes it possible for those with cars to maintain a given level of choice between facilities at ever-greater levels of dispersal. Conversely, its inconvenience for pedestrians, together with reduced perceptions of the safety of the network, act as further incentives towards increasing energy-consumption through the use of the private car.

At a smaller physical scale, the shift from compact terraced building masses towards individual, free-standing pavilions also has marked effects in increasing the consumption of fossil fuels such as oil and coal, because it generates an increase in the area of external building surface for a given floor

Transport Mode	Transport Type	(in MJ primary energy per passenger-km) Occupancy Rate		Type's Energy	Mode's Energy
		25%	100%		
Air	Boeing 727	5.78	1.45	254	254
Car	Gasoline < 1.4 litres	2.61	0.62	109	104.5
	Diesel < 1.4 litres	2.26	0.57	100	
Railways	High-Speed train 300kmh type: London–Paris	2.50	0.62	109	66
	Super sprinter	1.31	0.33	58	
	Inner City (standard commuter class urban rail system)	1.14	0.29	51	
	Suburban electrical line	1.05	0.26	46	
Bus/Car	Minibus	1.42	0.35	61	46.5
	Bus	1.17	0.29	51	
	Express car	0.95	0.25	44	
	Double decker	0.70	0.17	30	
Non motorised	Walking		0.16	28	104.5
	Cycling		0.06	10	

Figure 8.2 Different energy requirements for different transport modes.

area inside. For a given standard of insulation, this means that more energy is required to heat or to cool the buildings concerned, so as to maintain any given comfort level inside. And the production of this energy, in the form of electricity or heating oil, for example, uses yet more energy in its turn.

All these effects are exacerbated by the shift from locally-available building materials and components to those sourced in a global marketplace. For a given level of energy efficiency in the production of these commodities, the need to transport them over long distances means that this transformation leads to higher levels of energy consumption in the total building process. For a range of reasons, then, recent transformations have interacted with factors central to most users' cultures to generate dramatic increases in the consumption of energy, mostly through the direct or indirect use of fossil fuels. In turn, the burning of these fuels generates a range of atmospheric emissions, which reinforce the negative impact on air quality which we have already remarked.

In the light of all this, we can see that recent transformations at all physical scales are extremely negative in their effects on global air quality.

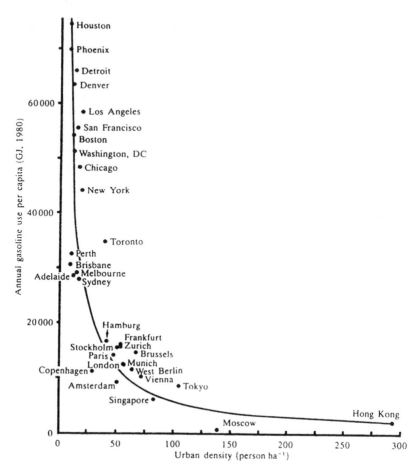

Figure 8.3 The relationship between density and petrol consumption per person.

The effects of the resulting toxic cocktail on any particular location, however, depend on its concentration in the atmosphere; and this in turn depends on the diluting effects of the flow of relatively uncontaminated air through the public space network. All things being equal, high rates of flow are easier to achieve in dispersed settlements, since there is less solid building mass to obstruct it. This means that the negative effects of dispersed settlement patterns are not immediately apparent through users' everyday experiences, and dispersion is likely to be perceived as improving air quality at the local level. Since it is a major cause of air quality problems in the first place, however, this is an extremely counter-productive perception in terms of users' own interests. If we are to work towards better-loved places, we have to find ways of reducing local levels of pollution which do not make the problem worse at the global level.

In denser settlements, which create lower overall levels of pollution per inhabitant, one key to reducing local pollution lies in the form of the public space network itself, which has an important impact on the speed with which pollution is diluted. As Bill Hillier and Pat O'Sullivan explain,[30] the rate at which air moves through the public space system of a given urban structure depends on what they call its 'smoothness' – the degree to which it forms a highly connected network, rather than containing stagnant enclaves. The transformation from grids to hierarchical layouts is all too obviously negative in these terms, and therefore leads to worse levels of local air pollution. Overall, then, the impact of recent transformations on the quality of urban air is highly negative.

Perhaps less obviously noticeable in the short term, but pervasive in its eventual results, is the impact of dispersed settlement forms on the general health of populations through the sedentary lifestyles which are encouraged when most travel has to take place through motorised modes. In Britain, for example, a national survey in 1992 found that about 70 per cent of men and 80 per cent of women take less exercise than is needed to maintain health,[31] whilst a 1990 study showed that people who cycle at least 25 miles per week halve the risk of heart disease by comparison with those who use only motorised transport.[32] Still more worrying in terms of what it portends for the future, however, is the finding that British children too get far too little exercise. A large majority, in the south of England areas studied by Neil Armstrong, took far less exercise than the minimum of twenty minutes brisk walking per week which would be required for health in later life, and Armstrong concludes that 'the current level and pattern of children's physical activity is a cause for grave concern.'[33] In sum, it is hard to avoid the conclusion that important aspects of recent transformations have had negative health impacts, and that these are likely to worsen as today's children grow up. But how have recent transformations impacted on the life-chances of non-human living things? How do these changes affect biotic support?

At the largest scale, the change from compact to dispersed settlement patterns, reaching out to replace wilderness with increasingly industrialised agricultural development, has clearly made life increasingly difficult for many non-human living things. In parallel, the increasing levels of pollution which are fostered by dispersed settlement patterns, impacting most directly on the densest urban areas, reduce the range of living things, such as native plant species, which can survive there too. In British cities, for example, the London Plane – first introduced into Britain about 1520 – has proved far more successful at coping with vehicle emissions than has the native oak. Not only does this increase the tendency to eliminate differences of local and regional identity, but it also has other ecological effects. The replacement of oaks with planes, for instance, has a major impact on the variety of insect species to be found in British urban areas: the oak, for example, has 284 insect species associated with it, whilst the plane has none.[34]

Similar reductions in biotic support also flow from the impact of global competition on particular localities, particularly through tourism, which seems set to become the world's largest industry in the not-too-distant future. Here many local biotic systems have been adversely affected by pressures to compete in the global market through globally attractive images. Here is a long-time visitor to the Australian resort of Noosa, north of Brisbane, lamenting one of the more noticeable biotic changes which have followed from the attempt to present Noosa as more 'tropical' by planting palms:

> The paperbark trees . . . and all those things which we rather loved have all gone. There are just remnants of all that bush. The birds in the street, because they had more natural trees, and now they have palms, and you don't have the same native birds.[35]

In many cases, however, the global reduction in biotic support is not so directly apparent in most users' everyday experience. After all, many compact settlements offer habitats only for the most aggressive and adaptable species, such as rats and cockroaches, which are often perceived in negative terms, whilst dispersed patterns allow for a wider range of habitats in gardens and also in spaces such as railway and motorway embankments, which have light levels of human use. Although recent transformations are negative in terms of biotic support at the global level, therefore, they may well be perceived as positive at the local level, by comparison with the compact forms which they supersede; and this perception is enthusiastically fostered by developers who draw on the importance of aesthetic issues in users' agendas to stress the green look of low-density developments with lots of 'landscaping'.

Given the arguments we have explored, we should expect that this green aesthetic would be highly appreciated by many users, and this expectation is amply borne out by surveys from many sources.[36] By the same token, however, we should expect that many other aspects of the aesthetic character of recent transformations would widely be regarded as negative, because their typical sensory patterns have come to be associated with types of settings whose impacts on other aspects of users' agendas have mostly been the negative ones. This alone may go a long way towards explaining why so many users have the aesthetic dislike of many recent transformations which we noted at the start of this book, but we must not allow ourselves to drift into a reductionist view of aesthetic experience which sees it only as a by-product of other cultural issues. This 'cognitive' level of aesthetic experience is important, but it must not blind us to the parallel importance of the 'perceptual' level, at which construction of aesthetic experience is affected less by cultural factors, and more by physiological ones.

At the perceptual level, we have explored the importance of richness in the pattern of sensory stimuli which any given setting offers, allowing the free play of physiological stimulus-processing systems, constructing aesthetic experience with a field bounded on the one hand by the chaos which results from trying to handle many stimuli at any given moment, and on the other by the boredom which comes from working with too few. How has the interplay between recent typological transformations and rules of good design worked through in practice, in richness terms?

So far as all the senses are concerned, a high proportion of any urban setting's sensory resources is generated through patterns of human activity. Viewed in this light, we can see that many recent transformations, across a wide range of morphological elements, have clearly reduced the range of resources available. As we have already seen, a simplified pattern of land uses in any given area follows from a range of interlocking transformations – from compact to dispersed settlement patterns, from a fine grain of mixed use towards single-use zones, and from highly-connected lattices towards hierarchical public space networks with more high-polarised levels of pedestrian flow. The upshot of all these mutually reinforcing transformations, all things being equal, is that the range of sounds, smells and visual events in any particular area takes on a simplified pattern. In most places the potential range of sensory resources is clearly reduced.

In itself, this is not necessarily negative in richness terms: at least chaotic sensory overload is unlikely, and it might be possible for designers with appropriate working practices to achieve richness through the skilful patterning of the relatively limited variety of resources which remain. Unfortunately, however, current working practices do not seem able to save the day in this regard. When attention is paid at all to richness at this larger physical scale, the efforts which are made all too often have the practical effect of reducing the variety of sensory resources still further. Even in the Netherlands, noted for advanced thinking in this area, the Managing Director of Amsterdam's Physical Planning Department has this to say:

> We recently discovered that . . . bakeries will no longer be permitted in residential areas. According to the environmental standards, bread smells bad. One of my colleagues recently commented that if this goes on, we shall be chopping down pine woods near residential areas because they don't comply with the odour standards.[37]

Overall, then, both typological transformations at the larger scale and the working practices of the designers who call on these types in particular projects conspire to reduce the richness of sensory resources in all but the visual realm. As we saw in Chapter 6, however, visual resources are extremely important to many designers; so, given the physiological under-pinnings of perceptual aesthetic experience, shared by people across the user–producer gap, we might expect that the results of designers' work here at least would be positive in users' terms. In practice, however, the interplay between recent typological transformations and designers' working practices does not seem to work out as well as we might hope.

As we saw in Chapter 6, it is primarily architects who are overtly concerned with the aesthetic dimension of design. For the most part, the architect's attention is focused at the scale of the particular building and its interface with public space. At this scale, recent transformations typically reduce the potential range of visual stimuli which given settings offer; the increasing size of building plots in an ideological climate of design 'consistency', and the shift from perimeter blocks to free-standing pavilion buildings with inactive 'backs' exposed to the public realm are both significant here. How well do current working practices allow architects to handle this reduced potential in the design of particular projects?

Unfortunately, these working practices do not work as well as they might. To begin with, as we saw, designers typically focus their efforts on the particular building with which they are concerned, without much consideration for those which form its perceptual context. Both in the visual representations which fuel the process of design itself, and in the presentation of the finished results of 'exemplary' buildings in the architectural press, the context is usually omitted altogether. In practice, therefore, each project is typically designed as an isolated 'set piece', which ignores the essentially serial nature of urban perception. The upshot is an essentially random relationship between each project and its serially experienced context. Whether the resulting relationships are rich, chaotic or boring is largely a matter of chance, rather than the result of careful aesthetic consideration.

This is not to say, of course, that only the relationship to its context matters when considering the aesthetic experience of any particular project. Clearly it is quite possible for users to focus their aesthetic attention on an individual building; and when they do so, the richness of its particular interface with public space takes on a discrete aesthetic importance. Even here, however, current working practices have a negative side. As we already saw, most building surfaces are designed with the aid of drawings at scales of around 1:50 or 1:100, generating visual representations which, at normal viewing distances, show how the surface concerned would look at a distance of some 25 to 50 metres. Assuming that the designer concerned can organise the design to have richness at this scale, this should result in a positive aesthetic experience at that range of distances. Unfortunately, however, most buildings in urban situations are experienced at far closer ranges than this, and some are also experienced from further away. The focus of the designer's concentration, in practice, is therefore on an arbitrary 'snapshot' of the building, again torn out of its serial context, at a distance which is often quite irrelevant to the range at which it is usually perceived. Whether the building is perceived as rich or not at any other distance is therefore merely left to chance: once again, the interplay between types and design practices is likely to be negative in richness terms much of the time.

Overall, then, the potential for richness is reduced by recent typological transformations, and this reduction is not well-compensated by aestheti-cally-orientated design practices, which all too often result in rich repre-sentations of projects, rather than rich settings in the world of everyday experience. Because aesthetic experience is in part related directly to the physiology of perception, rather than to social factors, this reduction in richness affects everyone. This matters; partly because aesthetic experience is important for its own sake, but also because the immediacy of perception means that aesthetic experience must be an important element in anyone's first impression of any given setting. If this immediate first impression is not an enjoyable one – as it will not be if the place is perceived as boring or chaotic – then users are likely to be 'turned off' from bothering to make the practical effort needed to get as much out of it as they might, in terms of their own life-projects.

To summarise, the cumulative impacts of recent transformation in terms of community, safety, nature, health and aesthetic experience do not seem positive from many users' points of view. Overall, we have seen that most people have been disadvantaged in terms of all these important aspects of the life-construction process; and whenever they have managed to avoid losing out, they have done so only through a dependency on resources such as the car which have negative implications from the ecological point of view. To take a positive stance, however, we should expect by the same token that there would be considerable common ground on which to build a powerful movement for change. As yet, however, we have not considered the implications which recent transformations have in terms of user-choice. If choice is the supreme quality, as our explorations suggest, then perhaps all the negative aspects we have so far discovered might be outweighed, in users' minds, if recent transformations could be seen as offering an enhanced choice of resources for users to draw on in constructing their life projects. Before we can map the common ground for change in a well-structured way, therefore, we have to explore from the choice perspective too.

9   Horizons of choice

We have argued that choice is 'the supreme value' for many users nowadays. If this is so, it must be important to understand how recent transformations in the form-production process, and in the forms of given settings themselves, have affected the choice of resources on which users can draw in carrying forward their life-projects.

As we have seen, mainstream ideologies equate choice in relation to the form-production process with the ability to choose one setting rather than another in the marketplace, through buying or renting. This is important to many users, but it is also important to remember that it offers very little choice in relation to the production of the public realm, for the simple reason that public space cannot be traded in the market. This means that the exercise of choice in the production of public space mostly has to take place at a political level. We should therefore expect that the potential for users to make their mark on the form and management of the public realm through some effective political process would be an important parameter in their evaluation of a place's quality, even where – like Bauman's Swedes – users might choose to exercise this choice by not taking advantage of this potential in practice.

The workings of government in representative democracies, however, are highly complex even at the local level. Even where politicians want to make themselves available to their constituents, this complexity makes it hard for them to do so. To many users, therefore, government processes seem like arcane mysteries rather than transparent, accessible systems. Even major businesses, with ample resources of money and skills, often find it advantageous to employ paid political lobbyists to gain access to the parts of the system they want to influence. To less powerful users, access to sources of political support is difficult indeed, as one very effective community leader explains:

> It's all like a maze. There's committees for this and committees for that, and you have to know who's the right person to contact all the time. And the language they use! It's all initials – DoE, SRB, all sorts. They're mostly nice enough people when you get through to them, but you'd think they were *trying* to keep you out![1]

Even where these problems can be overcome, so that users can gain access to sources of political support, the process of negotiating with the development team puts further hurdles in the way of effective involvement in the form-production process. For one thing the ability of users to make informed choices, even when they do get the chance, is all too often frustrated by the characteristics of professional cultures which we explored in Chapter 6. Most of the plans, sections and elevations which form the common stock in trade for architects, for example, seem totally opaque to most people outside the development team itself – certainly including most elected politicians. Unless they themselves happen to be design professionals, as is only rarely the case, none of the elected members of a planning committee, for example, is likely to have had any special training in designers' communication techniques, except what they learn on the job. In practice, their skills at decoding drawings are unlikely to be very much better than those of their constituents. They are advised, of course, by officials from the local authority's planning department; but in countries like Britain where the fragmented specialisation of the development team has gone furthest, these planning officials themselves will often have had only limited training in spatial matters, through a virtually 'form-free' education process. The following lament from a British development control planner, which I have read out at the start of many remedial training courses which I have run for others in the same field, always meets with rueful nods of recognition:

> As a development control officer with very little urban design training, I can offer only 'gut reaction' comment and advice which represents my own everyday experience. This tends to result in an unimaginative approach, rather like preparing an essay by copying out large chunks out of relevant literature.[2]

All this is compounded by a further dimension of the development team's overall fragmentation. Information about what is financially attainable, in practice, within the constraints and opportunities of the capital accumulation cycle, is usually restricted to developers and their financial advisers. Most professional planners, let alone lay members of the public or of planning committees, do not understand such matters in sufficient detail to know whether they can push developers towards schemes which users want, without pushing development away altogether. Unable to target their political power adroitly, their negotiating position is weakened still further.

In the most highly developed forms of the capitalist development process, then, there are few opportunities for most users to choose the kinds of urban places they want, so far as the public realm is concerned. But what about the concrete product of this development process? It is important to understand the implications which recent transformations have in terms of the amounts and types of resources which users have to deploy in order to maintain or improve the level of choice which they can draw from given settings. In practice, so we argued, the level of choice which users can wrest from given settings, with a given pattern of other resources to deploy, can be evaluated in terms of the key experiential qualities of variety, accessibility, robustness and legibility. How have recent transformations affected these qualities?

Let us begin by considering the ways in which recent transformations impact on the most fundamental support for users' choice: the variety of life-construction resources available in a given area. All things being equal, the variety of all kinds of resources in a given area has a strong relationship with the complexity of the land use pattern within it: the more complex the land use pattern, the greater the range of human contacts, goods and services available within the area concerned. Less directly the complexity of the land use pattern also generates the range of different forms of buildings and open spaces, and signs of habitation and use which in turn are linked to the level of sensory, aesthetic variety, in terms of the range of shapes, sounds, smells and so forth which are available in the area. The greater the complexity of the land use pattern, then, the greater the variety of resources of all kinds which can be drawn on to support all sorts of life-projects. The complexity of the land use pattern is therefore a key indicator of resource variety in general, and hence of the choice of life-construction resources available within any given area.

Viewed in these terms, the shift from complex, mixed-use patterns towards mono-functional zones, which we explored in Part II, is clearly a negative one. This transformation, as we saw, is fostered directly by considerations of profitability within the property development industry; but it is also locked in place by the cumulative effect of a range of other transformations whose impacts might perhaps be less obvious at first sight. To explore these impacts, we have to consider other factors besides direct financial pressures which affect the complexity of any area's land use pattern.

Amongst these factors, there are three whose importance stands out particularly clearly. The first is the density of human occupation of the area concerned: all things being equal, the more people there are in a given area, the more potential there is to develop exchanges and interactions of all kinds, whether social, economic or whatever. The second factor is the pattern of pedestrian flow to and through an area. The higher the pedestrian flow, again, the greater the opportunities to develop a wide range of social meeting places or economic opportunities: even the shops in the most car-orientated out-of-town shopping centres for example, ultimately depend for economic survival on the footfall past their windows. The third factor affecting the complexity of land use patterns is the level of rents which have to be paid by enterprises wishing to locate in the area: all things being equal, the higher the rents, the fewer kinds of enterprises can afford to be there. How have these factors been affected by recent transformations?

So far as densities are concerned, the transformation towards ever more dispersed settlement patterns clearly leads towards ever lower average densities, directly reducing the potential for a complex pattern of uses. In turn, this reduction is worsened by the shift from deformed grids towards hierarchical arrangements of the public space network, which leads directly to reduced levels of pedestrian flow in many of the spaces which form the network as a whole. Pedestrians, with only their own bodies' resources for movement, tend naturally to take whatever they perceive as the most direct route to wherever they are bound. In deciding which this is, they can only be guided by what they see from each point at which they find themselves. Pedestrians are therefore unlikely to go into streets which do not look as though they lead anywhere. This suggests that streets which are perceptually disjointed from one another, in the sense that the user cannot see far down them before the line of sight is obstructed by a change of direction, will be unlikely to be used as through-routes by people who do not live or have business in them; streets like this will therefore, all things being equal, have lower levels of pedestrian flow than others which are more directly integrated into the public space network as a whole.

This is amply borne out by a great deal of empirical evidence. Analysing the degree of integration in this sense, according to the pattern of sight lines (or axial lines, as he calls them), Bill Hillier shows that the pattern of integration itself usually accounts for over three-quarters of the pattern of pedestrian movement in the various parts of a given public space network.[3] The higher the level of spatial integration, the higher the level of pedestrian flow (again, all things being equal).

Now, the transformation of public space networks from grids to hierarchies has led to radical reductions in levels of spatial integration, as layouts have effectively become more maze-like. In turn, as we should expect, this has led to a marked reduction in the rate at which pedestrians encounter each other in public space. Comparing hierarchically planned, enclave 'estates' with more traditional grid-planned areas, Bill Hillier has this to say:

> In almost all estates, whatever the housing density, the overall encounter rate drops from an average of around 2.6 people per hundred metres/minute of walking time in street-based residential street areas (not counting shopping streets) to somewhere between 0.4 and 0.7 inside estates. The effect is that whereas in streets you are in contact with other people most of the time, on estates you are on your own in space most of the time. Put another way, the daytime encounter field in the estates turns out to be like night-time in ordinary urban streets. In terms of this naturally available encounter field, people on these estates live in a kind of perpetual night.[4]

The reduction in the number and variety of people in any given setting, brought about by these recent transformations, has direct effects on the complexity of land use patterns.

The viability of any non-residential enterprise, whether commercially or socially orientated, depends on the availability of enough people to support it. It is the presence of people passing through public space which provides the basic opportunity for human transactions of all kinds, amongst them the economic activities which are so important to the operation of any urban system. Pedestrian flow might be thought of as 'irrigating' the urban structure as a whole, providing the nourishment of human presence which creates the potential for capitalist enterprises to sell goods and services of all kinds. As population densities fall, the number of people needed to maintain any enterprise's viability must inevitably be spread over a larger geographical area. This in turn must mean that non-residential facilities themselves have to be spread ever further apart, so the grain of the land use pattern coarsens. In turn, this offers users fewer choices for a given level of mobility.

In capitalist situations, as we have seen, many resources are in practice made available through processes of market exchange. For a given level of disposable income, the choice of resources available in such a situation is therefore directly linked to the variety of capitalist enterprises with which particular users can interact. This is strongly affected by the transformation of public space networks from grids to hierarchies, which affects the levels of pedestrian flow through its various paths. All things being equal, high levels of pedestrian flow create the potential for high levels of economic activity. At the small scale, for example, this explains why the archetypal 'corner shop' is the *corner* shop, located so as to take advantage of pedestrian flows from both of the streets which meet there. At a larger scale, this relationship between the levels of pedestrian flow and the potential for economic activity is clearly demonstrated by the very direct relationship between pedestrian flow in any particular street and the levels of rent which can be achieved in the buildings which border it. Competition between enterprises which want to take advantage of the economic potential of locations with high levels of pedestrian flow pushes up the rents there; as we can clearly see from Figure 9.1, which shows how directly pedestrian flows and rents move in step along the length of Chicago's Ashland Avenue.

What this means is that there are very direct links between the level of integration between a particular street and the public space network as a whole, the pedestrian flow within it, and the pattern of land uses which this flow can support. These links can be seen very clearly in places where systems of planning control make little attempt to modify this pattern of uses. For instance, the city of Athens offers a rich source of examples which demonstrate how this relationship works in practice. In Figure 9.2, showing part of the Kipseli district, we can clearly see how differences of pedestrian flow, which follow from the different levels of integration of the various streets, in turn affect the pattern of uses. In this example, Patission Street – a major radial link into the city centre – is much the most highly integrated space. Of all the streets in the example, therefore, this is the one with the

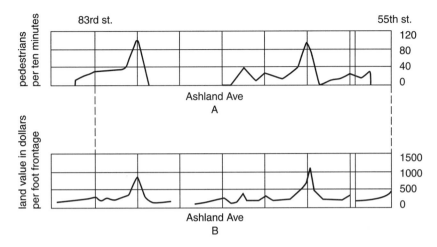

Figure 9.1 The relationship between rents and pedestrian flows.

highest levels of pedestrian flow. It is no surprise to see that Patission Street is almost continuously lined with shops, for these need high levels of pedestrian flow to survive.

The next highest levels of pedestrian flow, according to our argument, should be found in the streets such as Tinou and Ithakis, which are situated only one axial step away from Patission Street itself. This is borne out by the

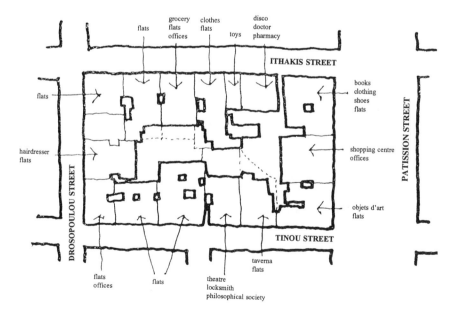

Figure 9.2 The relationship between spatial structure and patterns of activity.

fact that it is these streets which house the more specialised range of uses which people are more prepared to seek out, and which can therefore survive with lower levels of passing trade; I particularly like the combination of theatre, locksmith and philosophical society. Finally, we should expect to find the lowest levels of pedestrian flow in the least integrated streets – those such as Drosopoulou, which are positioned two axial steps away from the most integrated space of Patission Street itself. It is here, therefore, that we should expect to find the smallest proportion of commercial enterprises. As the example shows, this is indeed the case: this 'quieter' street creates conditions of pedestrian flow which can provide support only for housing.

To summarise, then, the shift from grids to hierarchical public space networks during the capitalist era is associated with reductions in the levels of pedestrian flows in ever-larger proportions of the network as a whole. This reduces the business opportunities available to capitalist enterprises, particularly for the commodity-sales outlets on which, in practice, the ability to choose resources so directly depends in most capitalist situations.

Variety is radically reduced by all this, and the situation is made even worse by transformations in the building stock which are brought about by the constant pressure to redevelop old buildings whenever a profit can be made by doing so. Even in areas where there is enough pedestrian flow to support non-residential uses, redevelopment leads to a reduction in the variety of uses – and therefore in the variety of all sorts of associated resources – because of its effects on rents. Except in particular situations where they have acquired cultural capital as 'antiques', old buildings can often be let more cheaply than new ones, both because they were built when construction and land costs were relatively low, and because they have already been earning money for a long time. In addition, there is often little competition between prosperous tenants to rent such buildings, because they neither match current standards of convenience nor boost the tenant's image.

Rents tend to rise when redevelopment takes place, for two main reasons. First, redevelopment makes higher rents possible, because new buildings can compete more effectively in the market, attracting a wider range of tenants, since they can be made to suit current space demands better than old buildings which were originally designed to accommodate earlier ways of doing things. Competing with one another in the capital accumulation process, building investors must obviously take advantage of the potential for higher rents if they are themselves to survive. Second, redevelopment also makes higher rents necessary if the investor's books are to balance. The costs of construction, and of borrowing the money needed to finance construction, together make it impossible for investors to charge the same rent after redevelopment as before it, if they are to survive in the competitive marketplace. For both these reasons, therefore, redevelopment raises rents, which in turn affects the pattern of uses in the area concerned. Many of the specialised shops, snack bars and so forth, whose presence contributes complexity and variety to the urban scene, can only afford to pay low rents – that is why they are usually located in old buildings in the first place. No matter how ideal a trading position it may be offered, this type of enterprise cannot afford to pay high rents. Even if given a free place in a redevelopment scheme it would almost certainly make economic sense for the enterprise to sell out and move elsewhere. Thus redevelopment, with its associated higher rents, reduces the variety of uses which can be accommodated. The ironmonger and the TV repairer give way to the jeweller and the boutique.

This reduction in the variety of capitalist enterprises directly reduces the choice of commodity-resources on which users can draw, even if they have the disposable income necessary to do so. The effects here are worsened by the fact that it also reduces the variety of resources in the aesthetic realm, since it directly reduces the potential for variety of sensory experiences. In the visual sphere, for instance, a reduced variety of land uses in a given setting is linked very directly to a reduced variety of forms, shapes, colours and textures there, since only a reduced range of building forms and other supports are needed. A reduction in the variety of non-visual sensory experiences also results, since the reduced range of activities also generates a reduced variety of sounds and smells.

Whether this reduction in the variety of sensory experience is seen in positive or negative terms in a particular instance depends on the interaction between the particular pattern of activities concerned and particular users' cultural rules and resources. Recent transformations, however, have removed even the potential for keeping some and removing others, for example through planning controls. They have largely been removed for reasons of profit, before users' values could be taken into account at all.

This reduction in the variety of sensory experience has been exacerbated by the way interfaces between buildings and public space have been transformed from active towards passive forms, with fewer 'leaks' of experience outwards through windows and doors. Crucial in generating this shift towards passive interfaces has been the shift away from buildings related together in perimeter blocks, presenting only their fronts to the public realm, towards free-standing pavilions entirely surrounded by public space, to which they present both fronts and backs. The problem here, from the point of view of aesthetic variety, is that backs by definition contain any development's most private activities, whatever these may be in any particular culture. It is inappropriate to carry on these private activities in situations where they are visible directly from public space – that, in practice, is what private means. This implies that wherever a space housing relatively private activities, whether indoor or outdoor, abuts directly on to the public realm, there has to be some kind of privacy barrier between them. Even if none is provided in the short term, we can be sure that something will be erected in due course. Now virtually all buildings, in virtually all cultures, contain some private activities; almost all of them, therefore, have backs. The shift towards pavilion buildings, entirely surrounded by public space, therefore means that some of this public space must be faced by backs. The privacy barriers which are necessary, in these situations, create increasing proportions of inactive, blank edges to public space – edges without windows or doors – as the transition from perimeter blocks to pavilions proceeds (Figure 9.3).

The variety of experiences available to those in public space is reduced as the proportion of active edges decreases, reducing the signs of life which were formerly contributed to the experience of public space through windows and doors. These passive walls now offer the user a reduced density of visual events, allusions and connotations, and this is exacerbated by the transformation from complex, hand-made façades towards simpler ones constructed from more standardised elements. It is then still further reinforced by the increasing frontages of plots and buildings, which make fewer subdivisions of each block to create visual variety along a given length of public space.

Overall, then, the transformations we have reviewed have had cumulative, pervasive effects in reducing the variety of resources of all sorts available for users' life-projects within a given area. In principle, this reduction might always be overcome through improved levels of accessibility: the reduced range of resources available in a given area might be expanded by drawing on resources from a larger area. How, then, have recent transformations affected the accessibility of the urban resource-system as a whole, for a minimum expenditure of mobility-resources?

Figure 9.3 The blank backs which result from 'pavilion' development: Birmingham Repertory Theatre.

Since it is the public space network which gives access to urban resources, the total range of resources made available through a given expenditure of mobility-resources is very much affected by the layout of the public space network itself. In practice, the choice of available resources depends on the number of alternative routes which can be taken through a given area within a given total travel distance, which is greatly affected by the typical transformation of the public space network from the deformed grid to the hierarchical system. The impacts here are most easily explored with the help of a simplified diagram. In Figure 9.4, two conceptual street layouts have

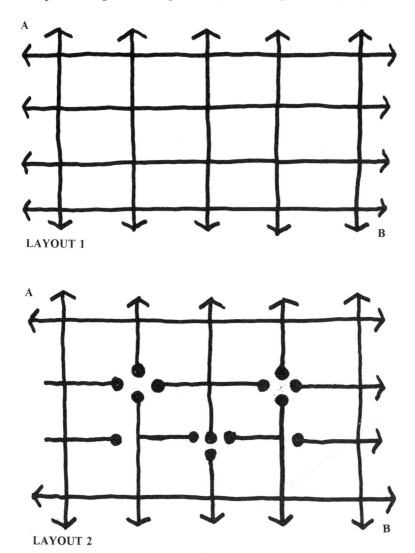

Figure 9.4 Choices, grids and hierarchies.

identical amounts of public space, but the systems of connection between the spaces are different. In Layout 1, all the spaces are joined together to form a grid. In Layout 2, in contrast, many of the spaces have been made into culs-de-sac, so that the system as a whole is far less continuously linked.

If we count up the number of alternative routes we could take without back-tracking, between points A and B which are in equivalent positions in the two layouts, we find a radical difference between the two systems: Layout 1 has thirty-six alternative routes, whilst Layout 2 has only two. This means that the two systems differ widely in terms of the variety of resources which they can make available to their users, even though roughly equal amounts of resources are invested in their construction. For an equivalent expenditure of resources in moving through the two systems between A and B, for example, Layout 1 offers eighteen times the choice of practicable routes, whilst some three-quarters of the length of the overall system in Layout 2 cannot be experienced – and can therefore offer its users nothing in the way of resources – without them 'going out of their way' (the everyday phrase is highly appropriate here) to make a special visit. If we generalise from this rather abstract example, we can see that the more any public space system resembles the highly connected grid of Layout 1, and the less it resembles the 'hierarchical' system of Layout 2, then the more choice of routes and therefore the more choice of resources it is likely to offer users who have a given level of mobility-resources.

The number of alternative practicable routes which the public space network offers to its users depends not only on the pattern of connections between its various paths, but also on the 'grain' of the system as a whole. If we compare the two highly connected grid layouts in Figure 9.5, for example, we can easily see that Layout 3 offers its users more alternative routes than does Layout 4. If the two layouts are drawn at the same scale, then Layout 3 offers only six practicable alternatives from A to B, as compared with the thirty-six made available by Layout 4. A similar coarsening of the grain of practicable routes through the urban fabric is brought about, in practice, by the transformation of public space networks from grids into hierchical forms, which effectively creates very large blocks of land which can be entered via culs-de-sac from their edges, but which present large-scale barriers to through-movement from one side to the other (Figure 9.6).

All things being equal, then, everything we have seen so far suggests rather strongly that the 'grid to hierarchy' transformation of the public space network, gradually building up throughout the capitalist era, leads to radical reductions in permeability, reducing the choice of routes which the network can offer, and with it the range of resources from which users can choose for the development of whatever life-projects they have. Far from recent transformations providing compensation for reduced variety through increased accessibility, our analysis shows that they compound the reduction in choice by reducing accessibility too. Users can only

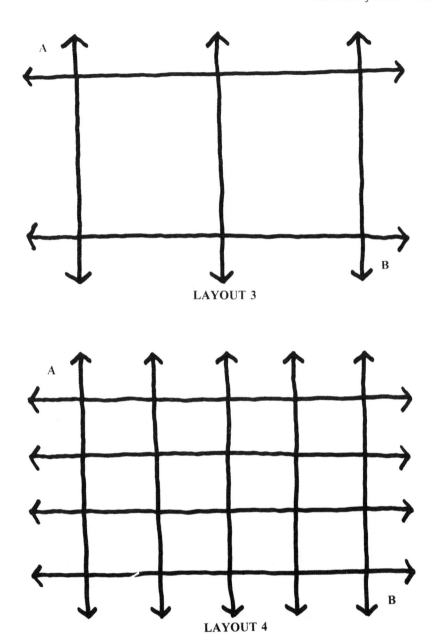

Figure 9.5 Choices and the 'grain' of grids.

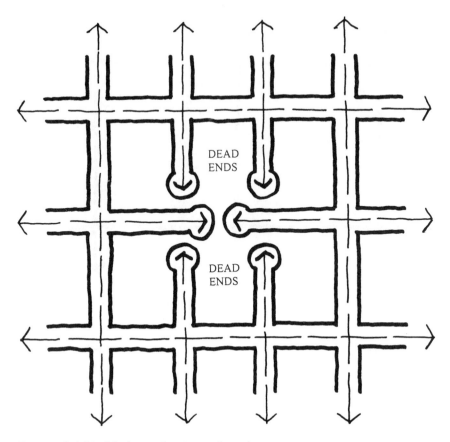

Figure 9.6 A big block as a barrier to through-movement.

overcome this double reduction, to regain previous levels of choice, by deploying extra mobility-resources.

   If such resources were available to all, perhaps the negative effects of the spatial transformations themselves would have little practical importance. Following this line of thought, it is often claimed that concerns about reduced levels of variety and accessibility are irrelevant today, because of the extent to which personal mobility has indeed increased in recent years. At first sight, this seems very plausible. It is certainly true that in the most 'advanced' countries, where the transformations we have discussed so far have been most evident, there have been dramatic increases in personal mobility, during the period when the transformations themselves have been going on. If we look more closely at how this has been achieved, however, we shall see serious flaws in the 'choice through mobility' claim. To start with, mobility has been achieved despite widespread reductions in the availability of public transport, which have themselves come about because more dispersed populations offer fewer opportunities for developing services which are financially viable without levels of public subsidy which prove hard to achieve in practice. In this situation, increased mobility has come about largely through the increasing use of private cars. But who has guaranteed access to the private car at all times?

To begin with, nobody is even legally allowed to drive a car until they reach the threshold of adulthood. Whatever age that is taken to be in any particular society, it certainly rules out those who count as children or adolescents in the society concerned. In other words, it rules out precisely those who, in the earliest stages of developing their life-projects, are most in need of the resources to take them forward. It also affects those at the other end of the age-spectrum: elderly people, who often feel less confident about driving as their eyesight, hearing and co-ordination worsen with age. This confidence reduction may become less significant in the future, as generations more wedded to car use grow old. Only about half of British retired households have cars, but the charity Help the Aged, for example, predicts that 90 per cent of men and 70 per cent of women in the 60–69 age group in Britain will have driving licences in 2021.[5] Even if they do not want to give up driving of their own accord, however, there are often government restrictions on older people's driving. In Britain, for example, people over seventy years old have to take supplementary driving tests at intervals, because of official safety concerns. With the 'grey' population growing throughout the industrialised world, a 'choice through mobility' strategy may well lead to progressive disadvantages for an ever-larger user-group.

Even amongst those who are entitled to drive and are capable of doing so, there are in practice marked disparities in terms of guaranteed access to cars. If we take the example of Britain, as a country which occupies a middling position in the league table of car availability, with an ownership rate per household far below that of the United States but somewhere between France and Japan, we can see how dramatically car ownership relates to household income.[6] Throughout the population as a whole, one-third of households have no car, a figure which rises to 40 per cent in Scotland and the North, where household incomes tend to be lower. If we take the bottom 60 per cent of households, in income terms, we find that in 1994, through Britain as a whole, almost half of them had no car. Of the bottom 20 per cent, the population without a car rises to two-thirds. In situations like this, it is clear that any strategy of 'choice through mobility' must have extremely reactionary implications, supporting those who already have the most resources, and making things worse for those with the least. Though it may now be weakening, there is also an important gender dimension to all this. In Britain, for example, only 35 per cent of women owned a car in 1996, though this was markedly up from the 25 per cent who owned one in 1985/86. Even now, though, many women in one-car households face what Jacqui Dunning calls 'the traditional convention that the male partner would have first use of the car for work and would always take the role of driver'.[7]

The numbing acceptance of the reduction in life-project resources which all too often arises from having to live in enclaves without high levels of personal mobility is poignantly caught, for example, in *The Growing Pains of Adrian Mole*, written, significantly enough, by a woman about the developing life-project of an adolescent boy. In the words of teenager Adrian himself, 'Sunday May 9th . . . I have just realised that I have never seen a dead body or a real female nipple. This is what comes of living in a cul-de-sac.'[8]

To summarise, then, we can see that most of the key transformations which given settings have typically undergone during the capitalist era have contributed to reducing levels of variety and accessibility, and therefore reducing the resources which given settings can offer for the development of most users' life-projects. In principle, these limitations might be overcome through increases in personal mobility; but we have seen that in practice many people do not have access to the resources they would need to make this work. Only those who are already in relatively privileged positions – largely young and middle-aged heads of households, in regular employment – seem able to stave off the choice-reducing effects of these transformations. Indeed, we might suspect that even people like these might not always find the transformations concerned totally congenial, since the very richest and most privileged people of all do not – emphatically *not* – usually live in suburban culs-de-sac. Even in Britain, with its supposedly anti-urban culture, the 'best addresses' are usually to be found in resolutely grid-planned areas.

Variety and accessibility, however, are not the only qualities of urban experience which matter. Recent transformations further reduce the choice of resources available to most users through their impact on the robustness of given settings – the potential which these settings offer for being used in a range of ways according to their users' particular desires. Both particular buildings and whole public space networks have mostly been given a reduced capacity to support robustness in recent times. At the level of any particular building, this reduction has been brought about through the shift from 'unspecific' types towards those tailored in advance for particular, predictable patterns of activity. The effects here are particularly pronounced when this 'functionalist' approach leads to the production of deep-plan buildings, relying on high levels of technology for their usability. Since the majority of human activities, in most cultures, require natural light and ventilation, this is a move which strikes at the heart of robust building forms. In terms of public space networks, a key problem lies in the fact that they have been progressively adapted to allow the free flow of motor traffic, used in attempts to overcome the effects of the reduced support for variety and accessibility which we have already explored. Traffic flowing at speeds of 30 mph, commonly the design speed for British and US networks even in residential areas, is highly dangerous to people who might want to carry out other activities in the space concerned.

The reduced capacity of given settings to support robustness matters most to those who are most limited in their capacity to travel about from one setting to another. Confined as they are, people in this situation need their limited geographical range to offer them as much choice as possible. The reduction in robustness which follows from recent transformations therefore has the most negative impacts on those users who are already seriously disadvantaged, in all the other ways we have explored, by a relative lack of mobility: children, elderly people, women and the poorest users in general. If public space is designed primarily as a channel for the convenience of the car user, then children's opportunities to play with their peers in the stimulating environment which public space potentially provides, for example, is limited by parents' fears about traffic safety. Similarly, the opportunities for less-mobile elderly people to relax in public space, to take advantage of the potential for informal face-to-face contacts which it offers, will be reduced by the noise and fumes of fast-moving traffic. Everyone except the car-driver suffers, as decisions about management and design, made to facilitate vehicular traffic, flush away the potential for other activities.

All the negative impacts we have explored so far, which follow from transformations which reduce the support for variety, accessibility and robustness, and which weigh disproportionately on the life-chances of those who are already most disadvantaged, are compounded by the way in which recent transformations affect the legibility of the urban scene, on which effective choice (rather than mere random encounter) partly depends. The importance which the layout of public space has, in terms of legibility, is clearly demonstrated by the research on mental maps which has been carried out by Kevin Lynch and his many followers. Analyses of maps drawn from memory, by a range of non-designer users, gives ample evidence of the importance which memories of the public space network play in their construction. In Figure 9.7, for example, we can clearly see that the 'skeleton' is made up of public spaces and their intersections – what Lynch, in his seminal *Image of the City*, called 'paths' and 'nodes'.[9] Lynch himself tells us, with reference to paths: 'For many people, these are the predominant elements in their image. People observe the city while moving through it, and along these paths the other environmental elements are arranged and related.'[10] It is not hard to understand why paths play such an important role in most users' mental maps. It is only the paths made available by the public space networks which enable people to use the city at all; so users who did not construct their mental maps around paths would find themselves completely at a loss in their everyday urban activities. The relative importance of nodes in many mental maps is again easily understood. Nodes, as the junctions of paths, represent points at which users have to make decisions about which way to go. Anyone unable to remember

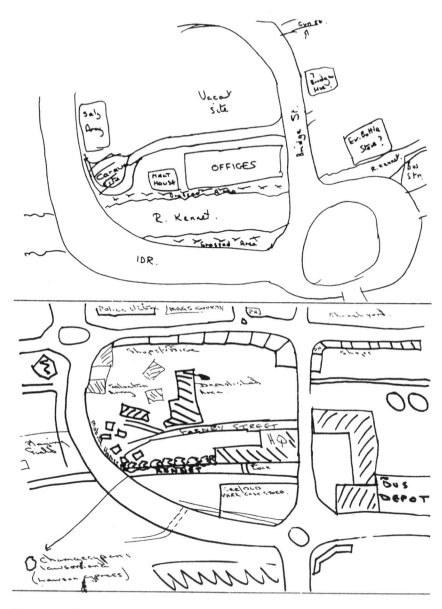

Figure 9.7 Mental maps.

these would only be able to use the city in a random, trial and error sort of way.

Potentially, then, the public space network offers important resources for the construction of users' mental maps. For this potential to be realised in practice, the public space network has to be formed to enable users to grasp local differences between different parts of the system as a whole. If different streets, for example, are difficult for users to distinguish, then it is easy to get lost at a local scale, as this resident of Jersey City tells us:

> It's much the same all over . . . it's more or less just common-ness. I mean, when I go up and down the streets, it's more or less the same thing – Newark Avenue, Jackson Avenue, Bergen Avenue. I mean, sometimes you can't decide which avenue you want to go on, because they're more or less just the same, there's nothing to differentiate them.[11]

On the other hand, however, if every part of the public space network were different from all the others, it would be an impossibly complex task to remember the *overall* pattern of the network, which is important for users in finding their way around at the larger scale. As well as local differentiation, then, a degree of global continuity is also an important resource for building up a usable mental map.[12] It is the *balance* between these local and global matters which counts. How has this balance been affected by the typical transformation of public space networks from deformed grids to hierarchical forms?

At the global level, the plan of any real deformed grid layout often looks chaotic, at first sight, to the casual observer; but this is not how it is experienced on the ground. In an analysis of Rothenburg ob der Tauber – surely an archetypal 'romantic' medieval town – Alan Alcock shows how the visitor never has to go through more than two axial steps to get from the edge of the city to its central facilities of cathedral and market square, along main connecting streets (Figure 9.8).[13] In turn, these main radial streets – acting like the spokes of a wheel – are directly linked by a 'rim' of major orbital streets, again arranged to give direct axial connections between one radial and another. It is this 'wheel' of main, direct connections which gives the settlement its spatial and perceptual continuity at the global scale. At the local scale, however, there is a great deal of differentiation between the various streets. As the plan shows, they vary widely in length, width and orientation, and in the angles at which they intersect. This local variation, however, is not achieved at the expense of dislocation from the larger, more direct 'wheel' on which the town's global continuity depends.

Even in the City of London – 'allegedly labyrinthine', as Hillier calls it[14] – the spatial layout within the 'labyrinth' is ordered in a way which maintains legibility at a high level (Figure 9.9). Hillier explains it like this:

> A closer look shows a principle. When after entering the complex you make a turn, thus losing sight of where you have come from, then either the second line already shows you another way out, or it takes you to an intersection with a line which does show you another way out. This

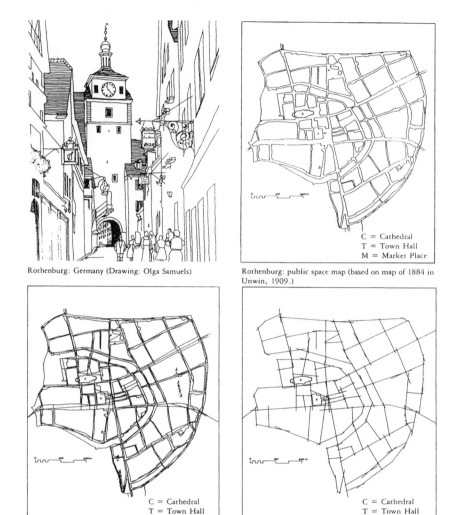

Rothenburg: Germany (Drawing: Olga Samuels)

Rothenburg: public space map (based on map of 1884 in Unwin, 1909.)

C = Cathedral
T = Town Hall
M = Market Place

Rothenburg: public space map showing axial lines

C = Cathedral
T = Town Hall
M = Market Place

Rothenburg: axial map

C = Cathedral
T = Town Hall
M = Market Place

Figure 9.8 Rothenburg ob der Tauber: a complex place with a simple spatial structure.

makes it difficult to go very far into the labyrinth. The line principle makes the complex easily intelligible, both in itself and from the main grid.[15]

This complex balance between local differentiation and global continuity, so typical of the pre-capitalist deformed grid, has been severely damaged by the transformation to hierarchical systems of public space. The axial map of

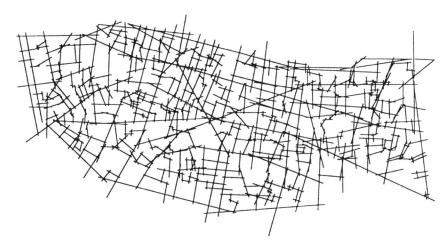

Figure 9.9 Axial analysis of part of the City of London.

part of Milton Keynes, for example (Figure 9.10), shows that this really *is* a labyrinth. Here we have at least three – and usually far more – turns to make from any point to get to the main grid roads – only to find, when we get there, that these have themselves been designed exclusively for motorised movement in any case. It seems clear from all this that the typical transformation of public space networks from grids to hierarchies reduces the resources which public space can offer for the construction of accurate and stable mental maps.

The problems of legibility which we all face in hierarchically-planned areas are made worse by the fact that the most complex of these are difficult to draw even on more formal maps, at any practicable scale. The *London A to Z* maps, for example, are drawn to the largest scale which can be fitted into a booklet of pocketable size. At this scale, it is possible to show all the streets in grid-planned areas reasonably clearly. But when it comes to the complex layouts of many recent social housing estates, like Angell Town, for instance, the system breaks down. All we are offered is an enigmatic gap, with 'Angell Town' written in.

Recent transformations at the smaller, plot development scale have also affected the legibility of urban places. Important here is the typical shift from connected masses towards individual free-standing buildings, which has led to a situation where public space itself becomes difficult to perceive as a positive element in the urban scene – as 'figure' – and begins to seem merely like a negative 'ground' of left-over space between figural buildings. This makes the public spaces of paths and nodes less memorable, and therefore less useful as resources for the construction of effective mental maps. As Lynch points out, what he calls '*Singularity* or figure-background clarity: sharpness of boundary . . . closure (as an enclosed square)' are

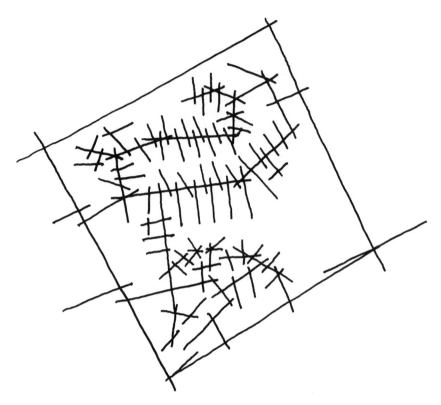

Figure 9.10 Axial analysis of part of the new town of Milton Keynes.

important amongst 'the qualities that identify an element, make it remarkable, noticeable, vivid, recognizable.'[16]

Given the importance which paths and nodes have in most people's
mental maps, the shift towards less memorable forms of public space is a
serious matter. This problem is compounded in those central areas which
have the most potential for capital accumulation through high-density
development. To achieve the required density of development, the free-
standing buildings here have to take on the forms of towers, outbidding
each other, in terms of height, to achieve their unique selling proposition in
the marketplace. In this situation, each individual building might itself
become a memorable landmark – that will certainly be its developer's
intention – but equally, in their juxtaposition, the buildings together are
always in danger of blurring into an overall district of similar forms: the
'central business district' as it is often called. In attempts to combat this
ever-present danger, each successive developer tries to achieve a building
whose image is still more memorable than those of its neighbours. Particularly in old-established city centres, this escalating process of dramatising

commercial space leads to a confusion of meaning. Given the associations made between size and importance in most cultures, the newer, tall and dramatised commercial buildings are likely to overshadow the importance of earlier and lower public buildings. Facilities which are of widespread importance, such as city halls, now look less important than buildings such as commercial offices, which are only important to a few (Figure 9.11). Important things look unimportant, and vice versa: a reversal of meanings which is confusing if not positively alienating in its effects.

Overall, then, it is clear that recent transformations tend to reduce the support for legibility which given settings can offer. The consequent difficulties which people experience in constructing either mental or externalised maps, can in practice have serious effects in reducing the resources which people can use in the development of their life-projects. For example, some of the British social housing estates where I have worked have been so illegible that residents have been quite unable to give prospective visitors intelligible instructions about how to find their homes. Residents of London's Aylesbury estate, for example, have told me that they prefer to meet visitors at nearby pubs, in adjoining areas which have legibly grid-planned layouts, and then escort them to their destinations. Clearly this presents people who are probably already disadvantaged in terms of income and mobility with yet other restrictions on their social lives, at the same time as it presents problems for others who have to work in such areas. Here, for example, is a comment from the person who has the job of meter-reader for the district heating system on part of Nottingham's St Ann's

Figure 9.11 Kuala Lumpur, Malaysia.

estate: 'The job's a nightmare at the moment . . . it's really hard finding the house you need to check.'[17] All this is very inconvenient, and even the most affluent of users cannot buy themselves immunity from these effects. Indeed, given their greater access to mobility-resources, they are probably the most likely to find themselves frequently in strange places, whose lack of legibility works against them.

The resources provided by the presence of outside visitors such as meter-readers are also important to those who are 'insiders' in any particular place. It is therefore clear that insiders, even though they may themselves have developed perfectly serviceable mental maps of the place concerned, through long familiarity, nevertheless suffer at one remove from outsiders' navigation problems. It is very irritating, no doubt, to have your social life disrupted or your mail go astray too often; but in the case of services such as doctors and ambulances the consequences of illegibility might be far more serious. Doctor Theodore Dalrymple, for example, gives us a hint of the problems he encounters in places with illegible networks of public space, particularly at night when there are no insiders around to offer directions:

> Take the housing estate I had just visited, a typical example of its genre . . . It was a meandering warren of roads, from which, once entered, it was impossible for a stranger to extricate himself unaided.[18]

Even local people could not help:

> Of course nobody – even those who had lived on the estate for years – had any idea where no.139a might be: certainly, it was nowhere near no.139. All my patients who require house calls seem to live at no.139a.[19]

Situations like this are made even worse because of the reduced pedestrian flows – and therefore the reduced opportunities for direct, face-to-face encounters – which result from the transformation from grids to hierarchical layouts of public space; as the Dutch architect Moshé Zwarts remarks:

> One clear day I was driving to the village of Soest, where I had an appointment. I lost my way and I could not stop, wind down my window and ask a passer-by the way. There simply were no pedestrians.[20]

Reductions in legibility, therefore, must surely lead to disadvantages for all users. Those who are intimately familiar with a particular place, able to build up their mental maps from memories of details which are not available to relative strangers, perhaps suffer the least. But given the transformation, for many people, to a more mobile lifestyle – itself made necessary, for example, by the parallel transformation towards increasingly dispersed settlement forms – we are nowadays all relative strangers in many of the urban places we use in the course of our everyday lives, and we are therefore all disadvantaged when legibility is reduced.

At this point, we have completed our review of the complex way in which recent transformations and working practice have affected the qualities which matter most in users' terms. Despite a few minor bright spots, we have found that the impact on experiences of choice has largely been as negative as it was with regard to community, safety, nature, health and art; and even where these negative effects have been staved off, this has largely been achieved only through ecologically negative levels of automobile-dependency. Though many people in the 'developed' world now have a greater choice of goods and services than ever before, urban settings themselves have largely been transformed in ways which do not make these choices available in the most positive ways.

For analytical convenience we have so far reviewed each transformation as though it had a separate impact on each experiential quality we have considered. In terms of users' real-life experiences, however, this is an artificial approach: the multiple qualitative impacts are all experienced *together*. Before we can use our insights in deciding how best to move towards better-loved places, therefore, we have to build up a better-integrated picture of the combined impacts which each transformation brings about, so as to develop a more comprehensive balance sheet of winners and losers overall. Let us conclude this part of our argument by pulling this balance sheet together.

Conclusion: an agenda for positive change

In the last three chapters, we set out to identify an agenda for change around which it might be possible to focus a widespread movement to get better-loved places on the ground, by exploiting the new potential opened up by the current sense of ecological crisis. We are now in a position, I think, to map this agenda out. There is certainly a depressing side to the picture we have uncovered, for it has become clear that there is a cruel contradiction at the heart of the capitalist development process. On the one hand, in the ideological realm, users are drawn towards particular kinds of values which enable them to make sense of life in capitalist situations. On the other hand, the need to survive in the competitive marketplace draws producers towards patterns of form and use which are largely negative in terms of these very same values, particularly for those who are already disadvantaged because of factors such as their age, gender, income or ethnicity. In the light of all this, it is not hard to see why recent places seem so often disliked – it would be surprising indeed if they were not. There is, however, a brighter side to what we have discovered. Our analysis has also enabled us to see what has to be changed, both in typological terms and in relation to the ways in which types are called on in design, if we are to get better-loved places on the ground.

At the largest physical scale, we have seen that the typological shift of overall settlement patterns from compact to dispersed physical forms, and from fine to coarse-grained patterns of land use, has many negative consequences. First, it reduces the choice of resources that users can employ in the construction of their life-projects, unless they have high levels of personal mobility. In the context of low-density dispersed settlements, high levels of personal mobility are difficult to achieve except through the medium of the private car, and we have seen the negative impacts which automobile-dependency has on air pollution and in fostering a sedentary style of life, and the negative implications these have for the health of many users. We have seen that the reduced encounter-rate between people in dispersed settlements diminishes the potential for forming communities of choice, and also reduces the sense of security which comes from the presence of other people deterring criminal acts through

increasing the chance of their detection. Not every impact of these transformations, however, is negative in users' terms. We have also seen that the
relatively large proportion of outdoor space which is under-used by humans,
in dispersed settlements, provides opportunities for planting and for urban
wildlife which increase users' contacts with nature.

Despite this particular bright spot, however, the existing picture here is
gloomy overall. This suggests that there is potentially a broadly-based
support for new types of overall settlement patterns – types which address
the problems of the ones we currently have. We need patterns which
provide high levels of choice when used through more benign forms of
movement such as walking, cycling and public transport, so that people
who can afford them can choose to use their cars when they want to, rather
than everyone being forced either to use them or to forgo the other choices
they want. We need patterns which provide users with the potential for
high levels of encounters with other people in public space when they want
this, and quieter spaces when they do not. And we want patterns which
provide these advantages without losing the potential for local manifestations of biotic support which current types provide.

At the level of the public space network, we have seen that the transformation from grids to hierarchies also has negative impacts in users' terms. It
reduces the variety of face-to-face contacts available in public space; and the
blank outsides of enclaves, presented to public space outside the enclaves
themselves, reduce both surveillance and visual richness. The advantages of
safety from fast-moving traffic, and for the expressions of imagined community which the shift towards enclaves provides, are both dearly bought;
there is surely a wide potential support for developing new types to achieve
these advantages without the negative impacts which they bring in their
train.

At the smaller scale of building form, the transformation from aggregated
masses towards free-standing pavilions surrounded by public space also has
many negative impacts. In terms of health, it increases energy consumption
and therefore atmospheric pollution. It makes it hard for public space to be
perceived as a positive 'figure' in the urban scene, and therefore reduces the
support it offers for constructing stable, accurate mental maps, or for
building strong conceptions of imagined communities. This suggests that
there is a wide potential support for new types which offer other ways of
promoting the saleable images of prestige which both developers and
purchasers require, but which do not incur these other disadvantages.

The relatively inactive 'backs' of pavilion development, exposed to the
public realm, also play an important part in deadening the interface
between buildings and public space. This transformation too has a range
of negative impacts in users' terms. First, it provides less potential for 'signs
of life' across the interface, with consequent reductions in the human
presence essential for constructing communities of choice and for surveillance alike. Second, it reduces the range of sensory experiences which

buildings make available to the users of public space; which in turn makes the built form less memorable, and therefore worsens the legibility of the overall urban scene. It seems clear from all this that there must be widespread support for a typological regime in which, so far as possible, public space never abuts the back of any building.

At the smallest morphological scale, further negative impacts flow from the transformation of surface materials from 'local' to 'global' in character, which contributes to a situation in which everywhere is seen as becoming more and more like everywhere else in visual terms, with a consequent weakening of the particular local or regional identity which might potentially give symbolic support for the construction of place-based imagined communities. There must be a strong potential support for new types to address this problem too. Finally, the transformation from complex to simple surfaces has a negative impact in the aesthetic realm, offering reduced opportunities for sensory choice. Virtually everyone, surely, wants to overcome this arid state of affairs: there must be widespread support for a typological vocabulary which will promote sensory richness, together with working practices which allocate enough design time to achieving it in practice.

Overall, then, we now have a reasonably clear agenda for typological change which might be widely supported at all morphological levels. It is also clear from our explorations, however, that a changed typological repertoire will not in itself be enough to get better-loved settings on the ground. For example, we have seen how the key desire for choice, as the supreme value in many users' agendas, is frustrated by professional ideologies of 'the expert' and 'the artist' in their current forms. And we have also come to understand how the increasingly global spread of these ideologies, in forms unresponsive to local users' particular aspirations, leads to further problems in many users' terms. The Expert and the Artist will both have to be reconstructed if they are to play the most positive roles they can, in promoting better settings.

We can understand, now, what has to change if we are to make positive use of the current sense of ecological crisis to achieve better-loved settings. Merely understanding what ought to be done, however, is a far cry from making it happen in concrete terms. How can these changes be brought about in practice? That is the subject of the next and final part of the book.

Part IV

Windows of opportunity

Introduction

Up to this point, we have been concerned only with analysis. We have focused entirely on why places have been transformed as they have, and why so many people feel disadvantaged by the settings which result. But this has only been a preparation for deciding how best to move towards better-loved places, which is what this last part of this book is about.

We have seen the potential for positive change which is being opened up, at the close of the millennium, by international political power in response to a widespread sense of ecological crisis. We have also seen, however, that these opportunities for change will not automatically lead to better-loved places, which can only come about in practice through concerted action by users and designers together. Given the producer–consumer gap as a structural feature of the speculative development process, users alone do not currently have either the 'seat at the table' nor the design skills they would need in order to influence the development process directly – users need designers on their side. Designers alone are in no better case; they lack the political power base they would need to buck the current system on their own, and they can therefore exert much more influence over events if they can gain widespread public support. Users and designers need each other's support before they can move forward.

As we saw in the last three chapters, users and designers share an important common ground of aims and values, on which a common movement for change might be founded. We also saw, however, that the practical impact of design culture's current mainstream repertoire of form-types and working practices is nonetheless negative so far as most users are concerned. Any move forward towards better-loved places, with more benign ecological effects, can only happen if this repertoire is radically changed.

From our structurationist perspective, however, we know that we cannot merely make up new types of forms and working practices from scratch; and even if we could, they would never come into good currency, in the mainstream of design culture, without the active collusion of designers themselves. Cultural structures such as types of forms or working practices are 'in designers' heads' as much as 'out there', so they can never be imposed, entirely from outside. If we want the mainstream of design culture to be occupied by new ideological material, positive in users' terms as well as in terms of designers' own values, then we have to bring about some subjective motivation, drawing both users and designers towards this new material.

This indispensable process of motivation has to operate at two levels simultaneously. At one level, this has to be a destructive process: the sense-making capacity of current mainstream design culture must be called into question by making it inescapably clear that current types and working practices have fundamental internal contradictions, because in practice they do not lead towards places which are 'good' in design culture's own acknowledged terms. I hope that the arguments of the last three chapters have helped to sow the seeds of this destructive process in at least some readers' minds. Destructiveness alone, however, will not do; if it is to lead to anything more positive than mere paralysing doubt, it has to be paralleled by a positive process. New cultural material must be made available, to enable designers to make a new, more credible sense of working towards better-loved places. In practical terms, time is of the essence so far as this double process of creative destruction is concerned, because of the sheer speed at which processes of urbanisation are currently happening around the world. It is no easy matter, however, to launch such a process of rapid, radical cultural change against the pervasive opposition which it must face from short-term economic interests. A great deal of moral commitment is required of any individual who hopes to resist these pressures. As we have seen, this kind of commitment is greatly strengthened if the person concerned has a sense of belonging to an imagined community of like-minded people, standing shoulder to shoulder within an established and well-rooted tradition.

As we saw in Chapter 6, most designers in practice still see themselves as members of the imagined community of Modernism. To an important extent, as we saw, the members of this community recognise one another through their shared use of particular aesthetic codes; but this is also a community with all the panoply of heroes, sacred books and the like which powerful traditions usually have. To ask designers to abandon the support this tradition offers is to ask a lot, and a programme of cultural change which makes this demand will appeal only to the brave and the foolhardy. If we are to build the mass movement for change which we need, therefore, we cannot realistically take this line. As far as possible, we have rather to seek our new typological repertoire of forms and working practices within the Modernist tradition itself.

Like all living traditions, Modernism is constantly constructed and reconstructed; and though much of what it currently contains is negative both in users' terms and in terms of its ecological implications, the overall Modernist tradition also contains far more positive potential. To reclaim Modernism as a resource for positive change, therefore, we have to identify these positive aspects, which constitute the best current raw material we can call on for working towards better-loved places.

If we are to address these problems effectively, we shall have to draw on this material to create both a new typological repertoire and new working practices for using this repertoire in particular design situations. Given the current cultural divorce between the social and the spatial – the split between 'product' and 'process' whose pernicious effects we have already explored – it is perhaps not surprising that most socially-concerned design thinkers have so far proposed cultural changes mostly at the 'process' level, calling for more widespread and effective systems of user-involvement, post-construction evaluation and so forth.

This perspective is clearly an important one, but we must not drift into considering built form itself in a reductionist way, feeling that the right product will somehow automatically emerge if the process itself is right. The logic of our argument so far suggests that any such comfortable assumption would be far from the truth. For example, the typological repertoire which designers put into play in any process of user-involvement, no matter how sensitively used, will itself have a very large impact on the socio-spatial agenda around which the design debate is organised. If this socio-spatial agenda is the wrong one from the point of view of many users, as is the case with the current mainstream typological repertoire, then the debate itself can only be fundamentally flawed from the user's perspective. In prospecting routes towards a more user-positive design culture, then, we also need to call on positive aspects of the Modernist tradition to develop a new typological repertoire, better attuned to the common ground of values which users and designers share. The outline of such a repertoire is sketched out in Chapter 10, 'Reclaiming the Modernist vision'.

To point to the need for a new typological repertoire, however, is by no means to diminish the importance of changing the process through which designers use it in their work. Certainly designers should not consider *any* repertoire as constituting some absolute, universally applicable ground of 'good design'. Real-life settings, after all, are never universal: they have to 'take place' – they cannot come into being without particular locations. And we should certainly not merely assume that any given typological repertoire, no matter how firmly rooted in the common ground we have explored, will necessarily be wholly relevant to the particular people who will use any particular setting. It becomes ever more important to avoid making assumptions of this kind as processes of global economic integration deepen, and the focus of the most intensive urban transformations shifts towards new geographical locations such as South East Asia. This shift leads to a situation in which job opportunities, for design professionals trained in 'advanced' countries, are increasingly to be found 'abroad', whilst increasing numbers of young designers now have the resources to come from 'abroad' to train in 'advanced' countries. Our current intake of graduate students at the Oxford Brookes University Joint Centre for Urban Design where I teach, for example, includes people from twenty-two different countries, all exposed – whether we like it or not – to the same Joint Centre culture.

Given these globalising pressures, it is urgently necessary to develop a design culture which refuses to assume that 'general' insights will relate to particular local situations in user-positive ways. We need a culture which assumes rather that the need to investigate whether they will relate usefully to local situations is itself a crucial part of the everyday design process. We need a culture, in other words, in which any typological repertoire we develop is seen in an 'open' way, as a set of potentially valuable hypotheses – to be taken not as universal truths, but as a useful agenda to stimulate fruitful interactions with particular local cultures.

As we saw already, however, the potential for such interactions is largely blocked, in practice, by various other aspects of the current mainstream cultures of design, with their generalised conceptions of 'the expert' and 'the artist', backed up by the notions of 'good design', supposed to apply everywhere, which we explored in Chapter 6. Potentially at least, this is one of professional culture's most serious shortcomings: whenever we call on generalised concepts of this kind, no matter how useful our insights might be in general, there is always the risk that we might relate them to particular local situations in counter-productive ways. If we are to make positive use of the insights we have developed so far, to benefit real particular users rather than idealised chimerae created as figments of professional imaginations, this state of affairs in professional culture will have to be changed. We shall have to identify positive raw material within Modernism, therefore, which we can call on to construct fruitful new visions of expertise and art around which new working practices can be formed. The potentials here are explored in Chapter 11, 'Experts who deliver', and Chapter 12, 'Artists in a common cause'.

By the end of this final part of the book, then, we shall have identified the windows of opportunity which we might open towards a new, more user-responsive design culture, which can use the new opportunities opened up by the current sense of ecological crisis to make effective impacts on the ground. Each of us, users and professionals alike, will be better able to decide which efforts we might make to push these windows open.

10 Reclaiming the Modernist vision

In Part III, we came to see why the experiential qualities of variety, accessibility, vitality, legibility, robustness, identity, cleanliness, biotic support and richness matter widely to users and designers alike; and we saw what was wrong with recent typological transformations in these terms, for most people, given the current patterns of resources which users can deploy. In criticising what was wrong, we implied a vision of what a better typological repertoire would be like. Now is the time to draw out these implicit insights, to identify the most user-positive strands within the Modern Movement as a whole, to construct a positive Modernist tradition to help us take advantage of today's new opportunities.

The typology we seek has to foster common ground between users and designers. This has important implications for how the typology itself should be expressed. So far as users are concerned, it has to be cast in a form which can readily be grasped without complex technical knowledge, but which encourages users to engage with fundamental design issues rather than superficial ones. From the designer's point of view, it has to leave as much scope as possible for the individual creativity which builds job-satisfaction, in order to engage the designer's active support for the new user-positive movement we seek. Taken together, these requirements have two key implications. First, they imply that we should specify only the irreducible minimum repertoire, restricting ourselves to types for each of the key morphological elements, and for the key relationships between these elements which are central to gaining high levels of support for the experiential qualities we seek. Second, it implies that we should specify the minimum range of types at as 'deep' and abstract a level as we can, so as to leave as much scope as possible for debate and for individual creativity in particular design situations.

At the most abstract level, we can see that any urban setting is shaped through some combination of human and natural factors. In capitalist situations (as in many others) we have seen that the ownership of land is a fundamental factor influencing the human forces at work, allowing particular agents to exercise power over particular sites, whilst prohibiting others. Crucial here, from our point of view, is the distinction between publicly owned and privately owned land – in concrete terms, the distinction between public space networks and development plots. To operate effectively in either the production or the consumption of urban space, at either the theoretical or the practical level, we therefore need a repertoire of types for these two basic morphological elements. These 'elemental' types are not all that matter, however. The qualities of urban settings are also very much affected by the typical relationships between these elements, and between both these and the 'natural' elements which are so widely valued. Our minimum repertoire of types, therefore, has to encompass these relationships too. What characteristics would these elemental and relational types have to possess, in order to offer the maximum support for the key experiential qualities we explored in Part III?

Let us start by considering the largest scale of morphological element. What should the public space network be like? As the basic spatial 'armature' around which urban form in general is disposed, the public space network has important impacts across the range of experiential qualities which matter in most people's lives. In terms of accessibility, it has to maximise access to as rich a range of life-construction resources as possible. In terms of public health, however, it has to make this accessibility available in a way which minimises pollution and which encourages the highest culturally acceptable level of beneficial exercise. Taken together, these requirements suggest that we need public space networks which offer high levels of accessibility without car dependency – public space networks which promote accessibility through easy pedestrian movement, cycling and public transport.

Accessibility is certainly not all that matters, however. Given the fundamental role it plays in the human experience of urban form, and also because of its longevity, a high level of robustness is essential too. This means, for example, that we should certainly not design so as to prohibit the use of private cars. After all, it is entirely possible that a practicable low-emission vehicle might be developed sometime in the future (though it is considerably less likely that everyone will have access to this). Our purpose, then, must not be to cut down choices for anyone, but rather to call on types of public space networks which increase the potential for achieving high levels of accessibility in clean and healthy ways, through walking and cycling.

An emphasis on walking and cycling, with public transport for longer trips, implies that a sense of safety will be crucial – particularly if adults are to allow children to make the pedestrian journeys which would have such a beneficial impact on public health in later years. This means that we have to find types of public space networks which offer the highest level of vitality, as indicated by levels of pedestrian flow for a given density of development within the area concerned. The desire to encourage walking and cycling also implies the need for a high level of legibility within the public space network, so that the youngest and oldest among us, with the particular orientation issues which they face, are not put off through the fear of getting lost.

Taken together, the need for accessibility through walking, cycling and public transport, for vitality, for legibility and for robustness strongly imply that we need a transformation from the hierarchical types of public space networks which are the current norm, towards better-integrated, more grid-like patterns. Highly interconnected grids, with many choices of routes between any two points, avoid the need for the complex, energy-wasting detours which so powerfully inhibit walking and cycling with hierarchical layouts, and which reduce the economic viability of public transport too.

In the short run at least, however, these changes may not actually reduce the amount of car use in practice. They will help to reduce people's dependency on the car, opening up opportunities for high levels of choice without it; but other cultural and economic factors come into play when people decide whether or not to take up these opportunities in their own everyday lives. As a worst case, therefore, we still have to consider which types of public space networks would offer the greatest degree of cleanliness for current levels of pollution generation. Here again, as we have seen, the grid layout scores over the hierarchy, because its 'open' connected structure allows the maximum clearance of polluted air for any given ambient wind effect.

Since they encourage pedestrian movement – as we have seen, some 80 per cent of the pedestrian flow within any public space can be predicted from its degree of integration into the overall network – highly connected public space networks also, by the same token, improve vitality. This has two interrelated effects, both positive in our terms. First, the increased rate of pedestrian encounters brings increased opportunities for the creation of communities of choice, whether of the face-to-face or the imagined variety. Second, the increased number of people around increases the scale of safety that users have. And because of the key role which paths play in the constructions of people's images of urban areas, the potential for long linking views which is generated through highly connected grids has a beneficial impact here as well. Finally, what about the robustness which is vital if the public space network is to create the possibility of 'open futures' for large-scale urban patterns? Here too, the grid scores heavily over the hierarchy for a number of interrelated reasons.

First, it is relatively easy to convert a grid into a hierarchy, if this becomes necessary for whatever reason. For example, there is a long tradition of converting grids into traffic hierarchies through management devices such as one-way streets, 'no entry' signs and the like. On the other hand, bitter experience shows that it is extremely difficult to convert a hierarchically-planned area, full of enclaves, into an open network structure without extensive problems of land acquisition and, often, resource-wasteful building demolition. Second, for a given length of public space, the open grid offers many more opportunities for connections to future extensions. This means that the issues of urban expansion which we face, both through population increase and through the increase in the formation of ever-smaller households, can be faced in more flexible ways, to address whatever may be the exigencies of ever more unpredictable futures.

Finally, the grid scores over the hierarchy in terms of robustness for a third reason: it allows each location within it to be reached by more than one route. This has a number of advantages. First, it makes even the most radical reorganisation of the pattern of uses or the building stock much easier to achieve, with the minimum degree of disruption. Second, it creates the potential for locating a wide variety of different land uses within a given area. The point here is that, in practice, objections to locating different uses in close proximity are often due to cultural factors far wider than the considerations of noise or other forms of nuisance which are often used to rationalise them. In many people's lives, it is the status differences between land uses which are really important here. For example, many people might object (in many UK cultures at least) to living in a house opposite a row of oily-looking workshops, no matter how quiet their operation, and no matter how useful the services or employment they might offer. If the workshops were 'round the corner' or 'round the back', out of sight unless one chose to go past them, however, the situation might be different indeed. The choice of routes which the permeable grid offers, therefore, is central to achieving the fine-grained pattern of different land uses on which the variety of all sorts of experience fundamentally depends.

Overall, then, quality considerations of all kinds overlap and build up towards the inescapable conclusion that the grid wins over the hierarchy hands down. We should therefore focus any limited power we can get hold of, to try to bring about transformations from the current hierarchical types of public space networks, towards more open, grid-like structures whenever we can. But what about the plots of development land thus created, which form our second key morphological elements? Which transformations should we seek here?

In urban situations, these plots are typically occupied by buildings, together with any associated outdoor spaces these might have. The key quality which enables these developments to have the most positive impact on the urban settings of which they form a part, in our terms, is robustness. This is important both for those who directly use the particular develop-ment itself, and also for those who use any larger urban setting which the development affects. So far as direct users are concerned, robustness allows them to choose alternative ways of putting the building to use in the course of their own particular life-projects. Users of the development's wider area of influence benefit less directly, in two ways. First, they benefit from the general improvement in resource efficiency which follows from the develop-ment's ability to cope with changing patterns of use over time: robust developments do not have to be wastefully torn down and rebuilt every time human purposes change. Second, they benefit because robustness enables the development to be used by a wide spectrum of different users, pursuing different activities, as its market rent falls in relative terms with age – a process which enables an increasingly complex pattern of land uses to locate in a given area over time, bringing with it all the advantages of experiential variety and vitality which we have already explored. Robustness, then, is crucial to plot developments which are positive in our terms. What would a robust plot development be like? In practice, it is the building part of the plot development which has to be considered most closely here.

Robustness of any building is powerfully affected by its height. Since it is the lower floors which are most easily accessible from the public space network which gives access to any building, it follows that floor-space at the upper levels generally appeals only to an ever-narrower range of potential occupants as the building's height increases; at least unless there are countervailing advantages in terms of prestige, views or whatever. This means that maximum robustness is usually achieved by building as low as possible, consistent with achieving the desired development density. The search for robustness also has important implications in terms of window-to-window depth, because the majority of potential occupiers will require natural light and ventilation. Buildings whose plan-forms are shallow enough to have all their floorspace naturally lit and ventilated, whatever dimensions that might imply in particular climatic and cultural situations, are therefore more robust than those with deeper formats.

To summarise, then, we have so far discussed types in relation to the two key human morphological elements – public space networks and development plots – which contribute the fundamental 'building blocks' from which urban settings are made. I have argued that the deformed grid type of public space network, together with the low-rise, shallow-built plot development which we have just discussed, are highly desirable because of the positive support they offer for the qualities which many users value. As we have seen, however, it is not only these 'elemental' types which matter within any typological repertoire as a whole. Just as important are the relationships between them, and between the ensemble they form and the natural elements which so many users value so much. In addition to the elemental types we have already explored, therefore, we need a set of 'relational' types which define the forms which these relationships should take.

When we consider the relationship between any particular development plot and the public space network, the key to making a successful setting in our terms lies in the amount of vitality which the development can contribute to the public realm, for this will affect people's perceptions of how safe the network is, and it will also influence the opportunities for the human contacts on which the capacity to construct both face-to-face and imagined communities of choice depend. To make the maximum contribution to the vitality of public space, each plot has to be developed so that it presents its most publicly relevant activities on ground level at the front, to create an active interface with the public realm; with more private activities at the back or on upper floors, to create the 'positive privacy gradient' indicated in Figure 10.1.

The active interface defines the most desirable type of relationship, from our point of view, between any particular plot and the public space which gives access to it. If this is consistently carried out, then a further relational type results: the ring-like perimeter block, with fronts facing outwards on to the public space network, and backs facing inwards towards the internal core of the block, as sketched in Figure 10.2.

The four types we have so far discussed – the deformed grid, the robust plot development, the active interface and the perimeter block – between them constitute a minimum, basic typological repertoire for supporting widely-sought experiential qualities. As yet, however, we have no type to define the best relationship between this ensemble and the natural morphological elements whose presence in urban settings is valued so widely.

Because it maximises the valued 'signs of nature' available through the everyday use of urban settings, there is widespread support for planting in the public space network itself. This is clearly positive in terms of the human values we have explored, but there are three further sets of relationships which have to be taken into account if this planting is to receive its full potential to support a rich biotic system with associated insects, birds and small mammals. First, this potential is enhanced by forming a spatially continuous biotic communication network, linking beyond any particular setting into other existing biotic systems, rather than merely planting for the aesthetic enhancement of particular public spaces, important though that is in its own right. Second, this network's potential for rich biotic development is further enhanced if it connects beyond the public realm, where it is subject to continuous human impacts and interferences, into the

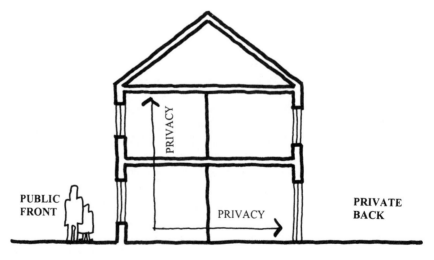

Figure 10.1 The positive privacy gradient.

Figure 10.2 The 'perimeter block' type.

least publicly-accessible parts of any given area, the backs of the perimeter blocks. Third, still further biotic richness can be encouraged by developing this network from native species, with deep roots in the local ecological system overall – a strategy which, by the same token, also reinforces the development of a distinctive local identity. The overlap between this complex web of relationships, stretching between the human and natural worlds, then creates the final type we need in our basic repertoire: the native biotic network (Figure 10.3).

To summarise, this overall repertoire, which I shall call the responsive city typology,[1] contains six basic types: the deformed grid, the complex use-pattern, the robust plot development, the positive privacy gradient, the perimeter block and the native biotic network. I am not, of course, suggesting that we should merely assume, without further thought, that this responsive city repertoire will be the best typology to use in every cultural situation; it is any creative designer's job to check this out on the ground in

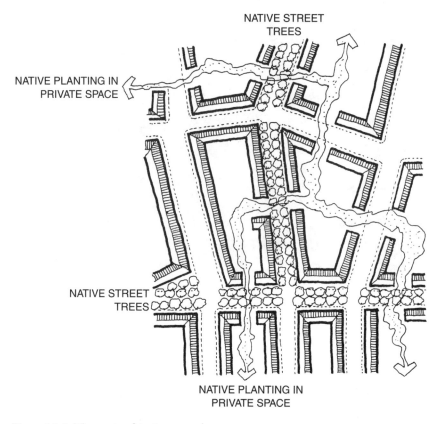

NATIVE STREET TREES

NATIVE PLANTING IN PRIVATE SPACE

NATIVE STREET TREES

NATIVE PLANTING IN PRIVATE SPACE

Figure 10.3 The native biotic network.

particular situations, and later we shall consider how this might best be done. It is interesting, however, that recent explorations into the urban design implications of 'non-Western' cultural traditions have suggested that the responsive city paradigm may have a very wide cultural relevance. Zain Abedin, for example, has found that this typology is in no sense counter to the tenets of Islam,[2] whilst Angela Ko has uncovered remarkable parallels between the responsive city approach and the tradition of Feng Shui, deeply embedded in Chinese culture and now widely influential across much of South East Asia.[3]

No matter how wide its cultural relevance, however, we must always use the responsive city typology in a culturally open way, as an agenda which can lead towards a wide variety of different designs, perhaps with radically different cultural inflections, in particular local situations. In this context, we have also to remember that the six types which constitute the responsive city approach are not the be-all and end-all of the typological repertoire we need to get better-loved places on the ground. We can use them, for example, only in relation to large-scale design issues, whilst in themselves they contain no aesthetic dimension. We have focused on these particular types because they can help us to address key problems which have arisen, in users' terms, as the Modernist tradition has developed; but that tradition as a whole is made up of myriad other, less problematic types which we still need in the practical process of real-world design. If we are to develop a practical fluency in handling these six types, therefore, we have to ask how they interact with the rest of the Modernist tradition; and we have to ask how users respond to the concrete designs which result from all this. Where should this exploration start?

Though the roots of the responsive city are very ancient, interpretations which we would nowadays call Modernist began to surface during the early years of the twentieth century, on both sides of the Atlantic. Depending on where they come from, these manifestations have quite different implications from our point of view. The United States, for example, had cheap land, cheap motor cars and cheap petrol which, together with the cultural value accorded to capitalism itself, allowed full rein to the development of dispersed urban patterns far earlier than in Europe. This is clearly negative from our 'sexy ecology' perspective, so most of the lessons we can learn from early US Modernism are concentrated at the scale of the individual building, where the strongly individualistic approach of many US designers can teach us much about the rich use of modern building materials.

When we seek lessons about the use of the responsive city repertoire at a larger scale, however, it is to Europe we must turn. Once again, we do best to focus on particular geographical areas: Germany and, in particular, the Netherlands. Here we find strong traditions of municipal town planning, together with powerful state and voluntary-sector programmes of housing production, to some extent removed from the direct pressures of the capital accumulation process. Taken together, these factors formed a working context within which creative designers could call on the types available at the time to develop a whole responsive city repertoire in new, Modernist ways.

To set the scene, let us begin by considering the 'Plan Zuid', the southern expansion scheme for Amsterdam (Figure 10.4). The plan was laid out by H. P. Berlage from 1902 to 1920, in an attempt to make city living more attractive to those people with high levels of disposable resources, who had shown a marked tendency to move out to lower-density surrounding areas.[4] Given that we face the same problem today, as we attempt to stem the tide of dispersion in favour of a sexy ecology, there is much we can learn from Plan Zuid, which is an object lesson in how to design with the whole vocabulary of user-positive types we have explored. A deformed grid of highly connected public spaces is defined by dense, adaptable five-storey buildings arranged in perimeter block format, with street fronts activated by entrances and windows to main rooms, with private backs as quiet urban oases. A mix of uses is integrated into this block structure, with commercial uses benefiting from the pedestrian flows along the most integrated streets,

Figure 10.4 Amsterdam's Nieuw Zuid plan.

and housing on the upper floors and in quieter locations, with lower levels of spatial integration. Uses which can contribute little to street vitality – the municipal tram depot, for example – are absorbed into the backs of perimeter blocks, but still with their entrances from the street, offering a first hint of how we might begin to deal with the 'dead sheds' of our own day. Finally, at the smallest scale, rich detailing ensures that the street level experience – in contrast to the famous aerial views – has great aesthetic interest. All in all, a textbook example of the deployment of the types whose value we have come to understand.

Over its three-quarter-century life, Amsterdam Zuid has also been warmly received by users and designers alike. Lay users have registered a positive vote with their feet and their rent books, as the historian Joseph Buch points out: 'it is the quality of Berlage's urban design that makes the area so desirable, and as the third generation of occupants move in, the average income level continues to rise dramatically.'[5] The place has been equally well-received, over many years, in the international mainstream culture of design. In 1934, when the area was still quite new, the US planner Catherine Bauer remarked that, it was in Holland that . . . Berlage and his followers . . . achieved the first real vernacular of modern architecture. This was particularly true in Amsterdam, where entire districts . . . bear witness to a fresh approach to the modern world.[6]

Over half a century later, eminent designers still see the Zuid area just as positively: Berlage's plan has now acquired an 'emblematic value', says Bernard Huet.[7] Plan Zuid, then, is a highly successful, straightforward interpretation of the responsive city repertoire. It has, however, a certain sober rectitude about it, which is by no means inherent in the responsive city approach. It is instructive, therefore, to compare Plan Zuid with other approaches which use the same repertoire in different ways.

In Hamburg, for example, Fritz Schumacher showed in 1910 how responsive city types could be called on to create designs with a far more complex range of settings (Figure 10.5). The scheme clearly demonstrates the great variety of spatial characteristics which this repertoire can support. We find a wide range of different spatial forms in particular parts of the scheme, but rather than being separate set pieces, they all contribute to the vitality of the whole; because each space maintains firm connections into the overall spatial network, thereby fostering the potential for people to walk through on their way to somewhere else, and also because the whole area is made up from the relational type of the perimeter block, so that all the public spaces, whatever their forms, are bordered by potentially active building fronts. What we find here offers us ways of addressing two of the issues which we have identified as central in today's urban design situation. First, it shows how we can achieve high levels of aesthetic richness, at a spatial level, without sacrificing the vitality and accessibility advantages of a highly linked spatial network. Here we can certainly have a rich sequence of different spatial experiences, but they are strung together like jewels on a necklace, so that the particular character of each space is not achieved at the expense of discontinuities in the whole. Second, the Hamburg layout shows us how particular public spaces can be designed with enough sense of distinct identity to foster a sense of imagined community, without erecting culs-de-sac which exclude everyone else. Not only does this maintain a high level of accessibility through the area, but it also offers symbolic material for constructing a subtler sense of imagined community than the symbolism of the cul-de-sac promotes. Instead of the cul-de-sac's exclusive 'us versus them', what we have here is the potential for a more inclusive, less divisive 'us *and* them'.

It is not only the spatial form of the public realm which affects the potential for users to construct an appropriate sense of imagined community: the surfaces of public space are important too. As we have seen, the 'everywhere the same' connotation of much Modernist imagery is often taken as negative in this regard, since it appears to offer no potential for constructing a sense of any particular 'us'. Certain early Modernist work, however, offers us ways beyond this situation.

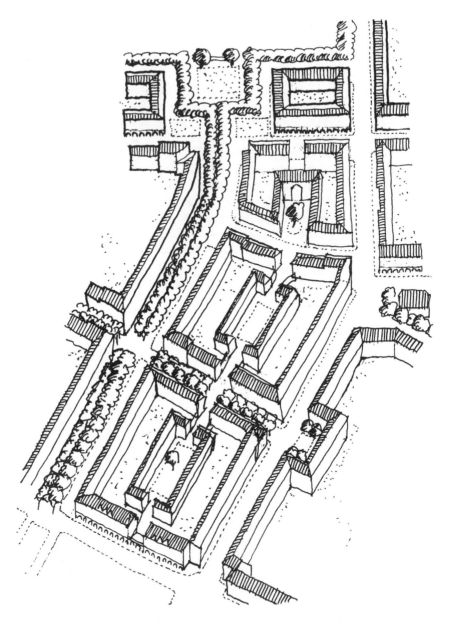

Figure 10.5 Complex use of the responsive city typology: Hamburg-Horn, 1911, Fritz Schumacher.

In the years before the First World War, the various constituent peoples of the Austro-Hungarian Empire, such as the Czechs, were pressing for independence, and the urban public realm was seen as an important medium for constructing a sense of 'us' to help promote this campaign.

Rather than merely seeking some revival of a supposed golden age before colonial oppression, however, Czech avant-garde designers sought forms which had connotations of both 'Czechness' *and* 'modernity'. The 'Czech Cubism' which resulted built bridges between a faceted style particular to Czech architecture and the latest, most 'modern' aesthetic experiments of Braque's and Picasso's 'analytical Cubism' from Paris. Rather than suggesting the xenophobic 'us-ness' of much nationalist art, therefore, Czech Cubism suggests, like Schumacher's spatial structures, a more inclusive 'us *and* everyone else' community sense, further supported by the way the individual buildings of this movement are carefully designed to emphasise a sense of continuity with the larger urban settings of which they form a part (see Figure 10.6).

Figure 10.6 Apartment building: Neklanova Street, Vyšehrad, Prague.

The Czech schemes were too small, however, to offer us further ideas about how the design of larger spatial structures might further contribute to this approach. If we return to Amsterdam, we can learn further lessons from the so-called 'Amsterdam School' schemes, particularly in the Zeeheldenbuurt area (Figure 10.7). Here, the structure of the public space network is again laid down as a deformed grid, but the form of the grid is manipulated in detail to create a great variety of integration levels between the various streets and squares within it. Levels of pedestrian flow, therefore, vary from busy to quiet, generating a rich range of settings, suitable for a wide variety of activities, within a small geographical compass.

Levels of pedestrian flow are highest, for example, on Oost Zaanstraat itself, which therefore supports various kinds of shops, a café and a post office. Only a few paces and one axial step away, however, the courtyard of Zaanhof forms a quiet green oasis, its 'apartness' emphasised by entrances through archways (Figure 10.8). This is certainly an enclave, but it is an enclave which forms an integral part of a well-connected street system, and it is entirely surrounded by active building fronts with entrances and with windows from the dwellings' main rooms; no public space here is 'round the back'. Again only a few paces away, but on the other side of busy Oost Zaanstraat, lies the large square of Zaandammerplein. As implied in Figure 10.7, this has an even lower level of natural through movement – though only a single axial step away from Oost Zaanstraat itself, it is far less directly

Figure 10.7 Plan of part of Zeeheldenbuurt.

Figure 10.8 The Zaanhof courtyard with its arched entrances.

connected into the other streets around it. The consequent lack of natural through movement, for those in cars as well as for pedestrians, enabled Zaandammerplein to be used for many years in conjunction with the adjacent school (now converted into a local community centre) which forms part of an adjoining perimeter block. Surrounded as it is by continuous building fronts, however, Zaandammerplein is clearly also a part of the general public realm, available to all citizens when school is out. This encourages a multiple use of space which helps to keep densities high without starving either the school or the local population of outside space, thereby creating an impression that the density is lower than it objectively is.

All in all, we can see that the layout of Zeeheldenbuurt is creatively manipulated so as to provide public spaces of many different characters, forming settings appropriate to a wide variety of activities, within a small geographical area. To be useful in practice, this complexity has to be legible to the people who live in the area. As well as achieving a highly legible spatial structure overall, through its highly connected character, the area also has much to teach us about how the design of particular buildings can contribute to urban legibility. The detailed design of the block, for example, contributes to the local legibility of the adjacent Zaanhof, providing a landmark vista from within its main square (Figure 10.9). To many Modernist designers (and even to me, to a degree) the fact that this spire

Figure 10.9 A memorable townscape relationship.

houses nothing in particular, and is purely there for its visual impact, seems to transgress the Modernist ideology of 'honesty' and to give the scheme an arbitrary quality which is reinforced by the idiosyncratic and surprising nature of much of its close-up detailing, which uses established types in an extremely free and creative way. Whatever the pros and cons of all this, however, it has certainly not stopped Zaanstraat becoming an icon of Modern Movement expressionism, and it may even have contributed to the fact that it is certainly loved by people who live in it, one of whom took the trouble to show me round its finer points, the last time I was there.

From the user's point of view, the fact that the architect de Klerk drew on the well-tried perimeter block format, incorporating a variety of building types to generate a fine grain of mixed uses has provided a basic user-positive framework, containing millennia of tacit wisdom, which is spiced-up rather than destroyed by its designer's youthful exuberance – incredibly, he was only in his twenties when this major work was produced. This typological vocabulary has once again created a place which gives all the signs of being loved by its users, and has stood the tests of use over time well enough to seem worth radical refurbishment some seventy years after it was built.

Yet more evidence of the breadth of different settings which can be achieved with this typology is given by J. J. P. Oud's housing of 1925–9, with associated church, in Rotterdam's Kiefhoek area – an interpretation in terms of stripped white surfaces and repetitive detailing, heavily influenced by de Stijl painting and sculpture (see Figure 10.10). As different from Berlage as it is from de Klerk, and built to a lower density than either, this still uses the deformed grid – beautifully linked into surrounding areas – and perimeter blocks, albeit with open corners, which present active building fronts to animate the public spaces. Only the church is given a pavilion format, appropriate to its special, landmark use – for those who believe, after all, sacred space has no back.

Kiefhoek too has been well-received over the years both by its residents and by the design establishment, to whom it has become a canonic Modernist work. Its place in international design culture was celebrated by its inclusion in the New York Museum of Modern Art's 1932 *International Style* exhibition, and publication first in the book of the show, and later in just about every historical review of Modernism ever since. Now a historic monument, its physical fabric had reached its sell-by date by the late 1980s. With a great deal of resident participation, however, it was lovingly rebuilt from the ground up, with new internal plans but keeping its original public space network and building elevations, demonstrating once again the robustness of the responsive city typology.

Kiefhoek rejects the spatial variety of the Amsterdam School in favour of a more basic, straightforward rendering of the responsive city repertoire, but this stripped-down character is by no means inherent in its 'International

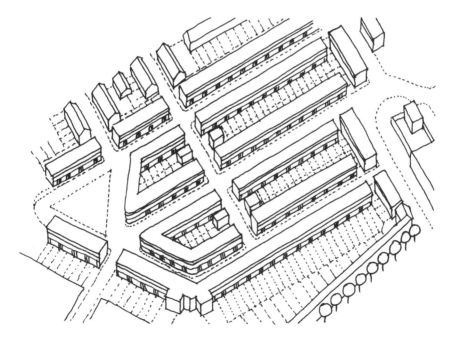

Figure 10.10 Kiefhoek: air view.

Style' approach. Amsterdam's Betondorp, for example, also shows once again the rich spatial potential of the typological vocabulary it uses, both in plan and in section. Noteworthy here, for example, is the handling of the housing scheme for elderly people adjacent to Brink, the central square (see Figure 10.11). Recognising the limitations of personal mobility which many older people experience, the housing here is located just a few steps away from the communal facilities of reading room, tenants' hall and shops which are arranged around Brink, on to which the dwellings look. To make the bustle of the main square less obtrusive in this situation, however, the housing for the elderly is slightly set back in an indent in the perimeter block, behind a light colonnade which doubles as a weather protected access to the communal rooms. To make sure that this arrangement does not encourage a sense of isolation, the set-back itself is used as a public garden, whilst the entrance porches into the buildings, each shared by four house-

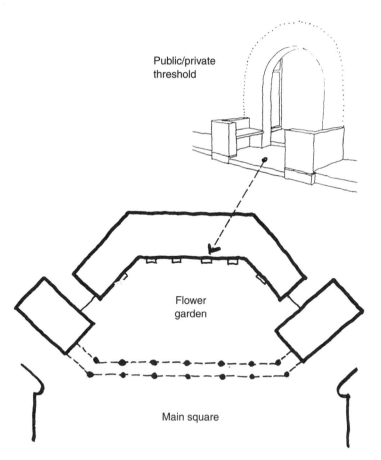

Figure 10.11 Seats at entrances, to encourage elderly residents to join in the life of the public realm.

holds, are generously scaled with small built-in benches to support informal contacts. Finally, at a smaller scale yet, rich details such as doors are achieved with thoroughly Modernist means (Figure 10.12).

The evidence of care shown in gardening and maintenance of private and communal spaces alike speaks volumes about users' positive reactions to this place, whilst the positive nature of current designers' attitudes comes over through both actions and words. At a tacit level, it is an architect's office which has elected to take over the old community hall, hit by the relative withdrawal of funding for the social facilities which has been such a marked feature of recent times. At a verbal level, the eminent architect Oriol Bohigas, for example, enthuses over 'the magnificent Betondorp'.[8] And again, these positive reactions – demonstrating the capacity of this resilient typology to cope with seventy years of rapid social change – have made widespread refurbishment seem a good investment.

At the largest scale, however, the relationship between Betondorp and the urban pattern as a whole has a more problematic quality from our point of view. Here we begin to see the impact of the urban dispersion whose economic rationale we explored in Part II: Betondorp is a relatively self-contained fragment, grafted onto the edge of the city as a whole. And it shows – as the architect Rob Krier puts it, 'the paths through the area lead to nothing.'[9] All in all, though, these three different performances of this common typological score stand up well to scrutiny from artist-designers themselves.[10]

Nevertheless, important elements of these places' typological vocabulary were gradually eliminated from mainstream international design culture in the years before the Second World War, through the material and ideological pressures and attractions which we discussed in Part II. At the smallest typological scale, a tradition of rich detailing was maintained, particularly in the United States through the work of Frank Lloyd Wright and his followers. But only in places at the margins of international design culture – places like Greece, Mexico and Venezuela whose small-scale development systems lacked the resources to overcome the practical logic of the grid and the perimeter block – was the capacity of a Modernist vocabulary to generate positive public spaces maintained after the 1930s.

We can learn valuable lessons, however, from the rare situations where the responsive city repertoire did manage to survive in the face of the capital-driven design culture ranged against it. One key example here is the large *quartiere* of social housing around Via Cavedone in Bologna, designed in 1959 with a grid street pattern and perimeter blocks, albeit with some of the street vitality reduced by too many blank ground floors of storage *cantinas*. I think it is significant that a prominent role in the design team here was played by Luigi Benevolo, also noted for his involvement in the contemporary public participation programme for the regeneration of Bologna's central area, which developed a methodology of urban analysis backed up by the inputs of the lay users, intended, amongst other things, to

Figure 10.12 A door at Betondorp.

uncover the tacit social logic of the street layout and block format of the pre-capitalist city. At Via Cavedone, I think, the legitimation which the experience of this process provides, allows the designer to exploit the fact that this is a social scheme without a profit motive, to buck the design-culture tide of the time; to produce a place which remains well-loved by those who live in it today.

Via Cavedone, however, is sadly the exception which proves the general rule of its time. Particularly for the most disadvantaged users, however, the problems of living with the larger-scale results of the destruction of this typological vocabulary, for example in large social housing estates, were beginning to make themselves felt in no uncertain manner by the 1960s, as Jane Jacobs trenchantly pointed out in her influential *Death and Life of Great American Cities*.[11] By the 1970s, these negative effects had become impossible to ignore, and the 1972 dynamiting of St Louis's award-winning Pruitt Igoe estate – an event whose symbolic impact on design culture was so great that the critic Charles Jencks sees it as marking the beginning of postmodern architecture – ushered in a period marked by the search for a more user-positive vocabulary of form.

With hindsight, this search can be understood as a series of attempts to learn from places which seemed to work in users' terms. Without much understanding of *how* they worked, however, particular types were often ripped out of context and used in counter-productive combinations. An interesting example is London's Angell Town estate, which had 'streets in the air' with far lower pedestrian flows than those which would have been channelled through a traditional street, bordered by blank walls (Figure 10.13). The problem here, in our terms, is that potentially user-positive elemental types are used without the relational types of the linked-up grid, the perimeter block and the active interface which hold them together to create positive emergent effects. As we shall see later, Angell Town was an unmitigated disaster so far as its residents were concerned, but I think it can also be viewed in a more positive light, as part of the painful process of action research which was needed to develop a practical understanding of the responsive city repertoire as a whole.

As long as this research process was confined within the desocialised culture of architecture or the despatialised culture of planning, without strong links to a popular culture in which the responsive city paradigm was a part of everyday lived experience, solid progress was difficult. It is no accident, therefore, that the first published explanations of these relational types – a crucial step in the typification process, as we have seen – came from the experience of professionals working with disadvantaged residents, who had no access to the resources they would have needed to overcome the problems of the capital-driven design vocabulary. From our perspective, the championing of the perimeter block by the Redevelopment and Housing Authority of Newark, NJ, in 1984 is an important step forward (Figure 10.14).

Figure 10.13 A pedway at Angell Town.

It is once again to Holland that we must turn, however, to see the full potential here, as institutions developed for user-involvement in urban regeneration set up well-structured frameworks for collaboration between professionals and users. This potential was seized on with enthusiasm by a generation of young designers in areas such as Amsterdam's Dapperbuurt.[12]

In the best of these schemes, strenuous attempts were made to allow as many residents as possible to stay in the areas concerned, if they wanted to. This meant that redevelopment had to proceed in a piecemeal way, retaining the basic well-linked grids and perimeter block arrangements which these renewal areas had usually inherited from the nineteenth century. What we see here, therefore, is a series of creative young designers attempting to update this inheritance into appropriate supports for the late twentieth century, with more spacious apartments, trees in the streets, new approaches to traffic calming, creative ways of achieving close-up visual interest, and often more active interfaces with balconies and steps coaxing signs of life from the apartments on to the streets themselves (Figure 10.15).

Places like Dapperbuurt have been around long enough for any signs of negative reception to show. The results, in practice, are overwhelmingly positive; and even the minor parts which have gone wrong bear out the

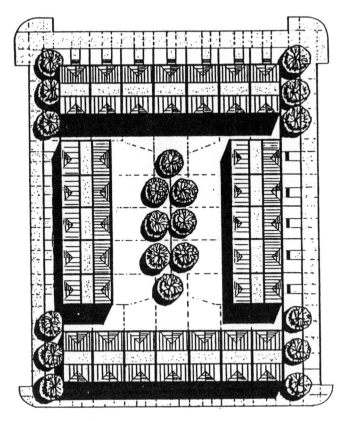

Figure 10.14 The rediscovery of the perimeter block by the Redevelopment and Housing Authority of Newark, N.J.

validity of the responsive city approach as a whole. The only area of Dapperbuurt with negative graffiti, for example, is a street which has signs of life only on one side, with a railway embankment blocking surveillance on the other. The negative effect this has, in vitality terms, is compounded by the unfortunate fact that some of the buildings themselves have relatively 'dead' interfaces with the street at this point, and by a serrated edge to the street, to adapt to the geometry of the site, which creates ideal 'lurking corners'. Since it violates these principles of the responsive city repertoire, it is no surprise to find that this is the least-loved corner within Dapperbuurt as a whole.

It is interesting to see that the degree of user-involvement in the design process which was achieved in the Dutch urban renewal schemes appears to have encouraged a finer grain of mixed use than would have been likely without it. When they are unable to check the responses of actual users, our own UK experience suggests that professionals are understandably, and for

Figure 10.15 Vitality and richness in a green street, Dapperbuurt.

all the right reasons, extremely concerned about the potential for non-housing uses to act as 'bad neighbours' in relation to the housing which inevitably makes up the larger part of the urban fabric as a whole. Wherever it is impossible to check this out, therefore, professional culture encourages a drift towards zoning, even when professionals are intellectually convinced that this would be negative. When prospective users' views can directly be canvassed, the situation is often quite different from that conjured up by these generalised fears. In our own work at Angell Town, for example, we found that elderly people had no objection to living over workshops, which they saw as potentially providing work for youngsters who might otherwise be hanging about the streets with nothing to do.

The capacity for high levels of user-participation to foster a fine grain of mixed uses, and the role which this can play in promoting the multiple use of public space, with all the advantages we reviewed in Amsterdam's Zeeheldenbuurt, is also amply demonstrated in the redevelopment of Rotterdam's Oude Westen. Here we find, for example, a combination of school, community centre and library near the city centre at Josephstraat, designed in 1982 (Figure 10.16). These three elements can be used either together or separately, whilst the play space forms the area's public square. As I walk along the street here, I feel that education is directly and symbolically linked with the life of the city – and vice versa – in a way which would be quite impossible in a more segregated situation. And the mutual surveillance of the square from the surrounding housing, café and community office, and the

Figure 10.16 A. van Wijk School, Oude Westen, Rotterdam, 1982.

complementary surveillance of the housing by the children when grown-ups are at work, seem to me an object lesson in how to make places feel simultaneously safe and lively.

The Dapperbuurt experience also suggests that the grain of mixed use attainable within a given culture also depends on the order in which the various elements of any mixed-use scheme are implemented. The fact that the noisy, messy but extremely lively street market was already in place before the adjoining housing, for example, meant that people who valued its vitality over its mess could be identified in advance, and offered the apartments which overlook it. If the market had been suggested after the housing was built and occupied, the chances of making it happen would probably have been extremely small. The responsive city repertoire, then, should not be thought of as offering designers a static snapshot of how a good city should be. Rather, it has to be considered as part of a wider process, operating through time.

Finally, in this account of the lessons which have been learned from the creative updating of earlier manifestations of the responsive city repertoire, it is useful to focus on Beijerlandse Laan, in Rotterdam's Feyenoord district. Originally designed in the 1930s as a major radial route into the city centre, this has now been rethought in a much richer, more complex way as a public space in which traffic has to co-exist sympathetically within other uses. The rethinking here was triggered by hard-nosed economic considerations: the shops which line both sides of the street had, by the 1980s, begun to suffer from out-of-town competition. Reducing the space available for moving vehicles, the municipality in effect made this wide street into a linear car park, with aisles which still allowed a slow-moving radial access into the city centre. Now visitors could park next to the shops, and pedestrian comfort was improved with wider pavements, covered with simple canopies. Soon the shopkeepers began to use them also for displaying their wares, and the image of the whole street changed from one of run-down decline to a bright lively richness. The upshot of all this, in economic terms, was startling. Interviews with the shopkeepers suggested that in some cases turnover had increased by up to 20 per cent, pushing them over the economic survival threshold in a most satisfactory way. So important were the canopies, in the shopkeepers' terms, that when the local authority found it had to demolish the original ones because they contained asbestos, the shopkeepers agreed to fund new, aesthetically richer ones out of their own pockets.

Beijerlandse Laan's lessons about how we might design streets to compete effectively with the dead sheds of the out-of-town mall are paralleled by other explanations of ways in which the dead sheds themselves might be integrated into areas structured through the responsive city typology. The problem here is that these sheds can themselves only have a maximum of 25 per cent of active edge if they are to work effectively for their internal purposes. Only this much, therefore, can be exposed to the public realm, if an active interface is to be maintained; so dead sheds have to be supplemented by other uses if they are to make user-positive settings. Effectively they have to be wrapped, wholly or partly,

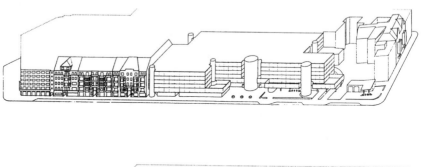

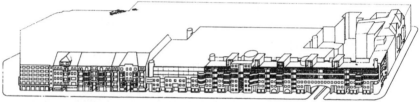

Figure 10.17 Housing wrapping a car park: Meinecke Strasse, Berlin; unbuilt project by James Stirling.

in other more 'active' uses, as Berlage did with his tram shed in 1920s Amsterdam.

An early influential rediscovery of this approach was James Stirling's project for housing animating a multi-storey car park in Berlin's Meinecke Strasse (Figure 10.17) – never built but widely published, and supported by all the influential prestige of a world-class star architect. Many later projects have shown how successful this 'wrapping' approach can be in practice, whilst our own explorations of ways of using large retail sheds to promote a lively mixed use area, in Longton, drew very conscious lessons from Beijerlandse Laan's wide, treed and canopied parking street (Figure 10.18).

By the time of the Longton project, the responsive city typology was definitely moving rapidly into the mainstream of design culture. In Holland, the building of the perimeter blocks of the Venserpolder scheme on the outskirts of Amsterdam by Carel Weber, one of the most influential

Figure 10.18 Integrating dead sheds: Baths Road, Longton, 1996.

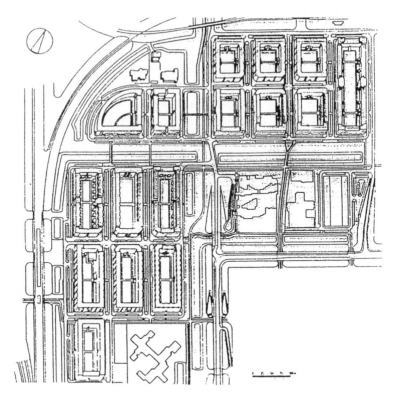

Figure 10.19 Perimeter blocks at Venserpolder, Amsterdam: Carel Weber.

Dutch architects of the time, gave added cultural capital (Figure 10.19), whilst the 1985 publication of our own *Responsive Environments*, now widely translated, no doubt helped to articulate a movement which was anyway waiting to be born. *Responsive Environments*, for example, has proved the most influential 'academic' text used in developing UK local plans, and was paralleled, with the involvement of users, in preparing the design guide for the redevelopment of the Hulme estate in Manchester, itself a 1960s development which had broken all the rules of the responsive city paradigm, to establish principles which were later expanded into a design guide for the city as a whole (Figures 10.20 and 10.21). And in Holland, where this account began, creative examples of the responsive city approach now abound.

All in all, then, this chapter's story is an encouraging one. Having been elbowed to the margins of design culture by the pressures towards profitable development, the responsive city repertoire is now being brought back into the centre ground, and transformed in creative ways to face up to the

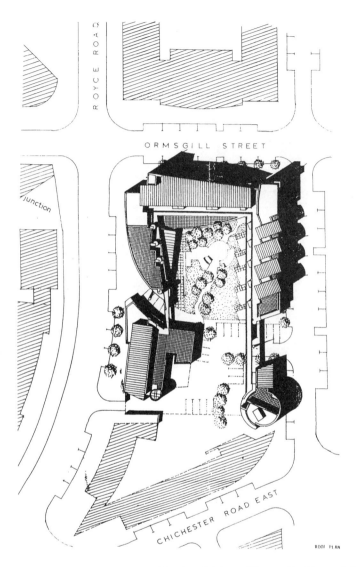

Figure 10.20 A lively perimeter block at Hulme: 'Homes for Change', Leavey Mills
Beaumont, 1994.

problems of our day, which the discredited current mainstream seems so
signally incapable of doing. As we already saw, however, the responsive city
typology is not a static product. Rather it is a set of cultural rules to guide a
dynamic, creative process through which the new opportunity-space opened
up by the current sense of ecological crises can be colonised to get better-
loved, more sustainable places on the ground. These places lie in the future,

Figure 10.21 Modernist perimeter blocks at Hulme.

not the past: if we want to achieve them, we have to work for them, with all the expertise and all the artistic creativity we can muster. There is a problem here, however: our current concepts of expertise and art, as we saw, are themselves mired in user-negative modes. The final two chapters will explore ways of transforming them too.

11 Experts who deliver

In Part II we came to see which aspects of mainstream professional expertise support the unloved status quo. Two of these aspects are particularly negative in our terms, because they make it difficult to move better-loved types and co-creative working methods into the mainstream of design culture. First, professional expertise has been developed with an eye to innovation, to enable those who have this expertise to steal a march on their competitors in the market for professional services; and this has led to a state of 'neophilia', in which innovation is commonly regarded as admirable in itself. As we saw in Chapter 3, this makes it difficult for designers to acknowledge the validity of any typological approach to their work. Further, it suggests that the types we proposed in Chapter 10, as currently offering the best cultural raw material for making better places, are nostalgic and old-fashioned because they have been around a long time, and should therefore not be on any expert's agenda. Second, professional work is increasingly structured through an ever-widening range of specialised disciplines, each related to some particular, limited aspect of design work. By defining only those involved in these disciplines as experts, professionals can feel justified in overriding the values and opinions of lay users – since users themselves are seen as lacking expertise, their views can legitimately be ignored. A myopic fragmentation rules, and the whole becomes daily more opaque to lay people's understanding.

The fact that there are aspects of current expert culture which are negative from our point of view does not, of course, mean that the idea of 'the expert' either could or should be abandoned altogether. The practical benefits of specialised, expert knowledge are manifest all around us, and are as obvious as any negative effects it might have. The problem, therefore, lies not so much in the fact that designers feel that they are experts, as in the kinds of experts they feel themselves to be. To foster the practical changes we seek, we need a rethought and refelt conception of expertise, which acknowledges the role of types in design, and which welcomes users into the process of design as allies, rather than excluding them from it. In this chapter, we shall explore ways of reimagining 'the expert' which are positive in these terms, calling on the opportunities which can be opened up by identifying and exploiting contradictions within the current neophiliac and myopic conceptions of expertise itself. We shall see that these current conceptions are internally contradictory, in the sense that they lead to practical results which are negative in the expert's own terms.

So far as neophilia is concerned, it is important to realise at the outset that any design approach which regards innovation as positive merely for its own sake, with no 'outside' indicator of the quality of the practical outcome, has a self-defeating circularity about it. The internal contradictions here have been pointed out many times, as when Jennifer Bloomer reminds us that, 'The ceaseless search for the New in Architecture . . . is a profoundly nostalgic project.'[1] She goes on to comment that: 'the legitimation and substantiation of the desire to be New has become so old hat as to form a bizzare paradox sustained by the now rather rickety crutch that is called Progress.'[2] In the end, then, it is nonsensical to take innovation in itself as a touchstone of design quality. If it is maintained as a touchstone over time, then eventually this attitude itself must become – indeed, *has* very definitely become – a tradition in itself. And once this particular 'tradition of the new' has become established, the only way of still being innovative, and therefore producing good designs, must be to break free from the confines of this tradition. But the only way of breaking free, to be original, is now by being *un-innovative*. A lack of innovation, therefore, becomes highly innovative, and only bad design is good – a self-defeating muddle quite counter to the ethos of coherence which underpins the very concept of professional expertise.

At a more practical level, the problems of neophilia arise from the fact that any innovation-focused approach to design tends to generate unpredictable results. This problem does not arise only when an innovative approach is taken to certain kinds of issues. Rather, the social reception of innovations of any kind appears to be inherently unpredictable – to the extent, for example, that despite increasingly sophisticated techniques of market research, current estimates of the market failure rate of new products currently range around 80 to 90 per cent.[3] Against this general background of innovation-failure, the widespread popular rejection of recent urban places seems almost a success – after all, only some two-thirds of respondents in a MORI poll put modern architecture into the 'eyesore' category.[4] Unfortunately, however, this lower-than-average failure rate cannot compensate for the fact that failures in given settings just matter more to most people than failures in terms of other products. If some new kind of gadget fails in the marketplace, only its manufacturer, its shareholders and perhaps their direct employees suffer. Urban settings are different: they have captive audiences, they affect more people over a far longer timespan, and if things go wrong they involve a far greater waste of resources.

None of this means, however, that innovation should (or could) be eschewed in all circumstances; since each new project necessarily has some unique aspect, some degree of innovation is always inevitable. Indeed it is also, to some extent, desirable: it is the search for innovation – the exercise of creativity as the 'supreme value' of choice in the work-process – which gives designers the high level of job satisfaction which we noted in Chapter 2. We shall not succeed in persuading designers to abandon the negative aspects of their current ways by suggesting that they should adopt a more boring approach to their work. In any case, even if we could succeed in such a move, users would not thank us: to a degree, after all, a constant process of innovation is a necessary prerequisite for providing urban settings with the variety and richness which users also value. Nonetheless, all innovations are to some extent unpredictable in their practical impacts, so we have to accept that unpredictability is endemic to real-world design. It is the degree of unpredictability which matters in practice, however; and an unfocused, neophiliac drive towards innovation in all aspects of design is likely to increase it, often with negative results. The concrete example of the Angell Town estate may help to bring this home. Since the estate was built as a social facility, rather than to make profits in the marketplace, the various professionals involved in its design were in principle released from direct pressures to support capital accumulation. Though there were definite cost limits, the designers had considerable freedom to decide the particular vocabulary of forms on which the available resources should be spent. Relatively free from profit-making pressures, this was a situation in which the designers had considerable freedom to work towards the production of a better-loved place.

Drawing on their professional vocabulary of types, and seeking innovative solutions, the designers called on various ideas which had recently come into good currency within design culture, with a particular view to improving the residents' safety by keeping cars and pedestrians apart. Cars were to run at ground level, whilst people on foot were supposed to move around the estate on pedestrian decks or 'pedways', linked from block to block by bridges, forming a segregated pedestrian network at an upper level; with many of the buildings' ground floors given over to garage space. As a response to the particular problem of conflict between pedestrians and vehicular traffic, this had an elegant technical logic which was very convincing to many designers of the time. Unfortunately, however, this typology was too new to have stood the test of time in use, and in practice it turned out that the physical forms which were designed for the purpose of keeping traffic and pedestrians apart also had other, unexpected consequences in terms of their users' total experience of the place.

First, access up to the pedways could only be provided through stairs and ramps which double back on themselves, with complex shifts of axis, so the pedways themselves are only weakly integrated into the spatial structure of the estate as a whole. This means that they are never used as short-cuts by people who want to go through the estate on their way to somewhere else, so the only people to be seen in them are those who have to use them for access to their own dwellings. These, in turn, are relatively few in number;

to allow light and ventilation to the pedways themselves, they could only have dwellings along one side. All this reduces the potential for encountering other people on any particular pedway – an effect which is worsened by the character of the interface between the pedways and the dwellings which open off them. For reasons of practical economics and construction, the pedways are of restricted width, and run directly past the rooms in the adjoining flats at the same floor level. To ensure privacy in this situation, the front walls of the dwellings are designed without windows, which means that the people using the pedways gain no signs of life from the people inside, as we saw in the last chapter. For those who react to this state of affairs by deciding to walk as far as they can at ground level, before climbing up to the pedway system, the situation is no better. Many of the buildings' ground floors are given over to garaging, which again contributes no signs of life to the adjoining public space, particularly since the lack of surveillance leads to a situation where nobody who values their car, if they had one, would choose to park it there in any case. Left empty, vandalised and derelict, the garages worsen still further the public space experience.

From the beginning, these physical arrangements, with their marked lack of surveillance, gave no support to community safety in a situation where such factors as low incomes and high levels of unemployment gave rise to considerable social stress. Many people began to feel that Angell Town was an unsafe place, and sensational newspaper articles made this perception worse: 'Angell Town becomes Hell Town', said a headline in the *South London Press*. In a survey carried out as part of a programme for improving the situation, it was discovered that some 80 per cent of women residents, for example, felt 'unsafe' or 'very unsafe' when they entered the estate from the areas around it.[5] The experts' innovations, designed to improve safety, had in practice contributed to the opposite effect.

What we see here is a chain reaction of unintended consequences which, with the benefit of a hindsight not available to the original designers, we can understand as flowing inexorably from the innovative approach to design – dumping the tacit knowledge embedded in the area's previous arrangements of well-linked streets and perimeter blocks, in favour of new forms. The creative, innovative application of expert knowledge here has clearly contributed to an outcome which can only be seen as negative, even according to expert criteria of 'efficiency' – to the extent that some £60 million is now having to be spent to put right an estate which had originally cost £27 million only twenty years before. Quite apart from its negative impacts on users' lives, this clearly represents a highly inefficient and unsustainable use of scarce resources.

The problems of unpredictability which must flow from any innovation-driven approach to design are often compounded, in practice, by the myopia which is associated with the specialisation of the design professions. Viewed positively this specialisation, structured through the complex professional division of labour, can be seen as a necessary response to the manifest complexity of urban places. If we are to think about this complexity in any depth, we have to concentrate on some particular aspect of the place – its pattern of uses, say, or its traffic flow – by putting conceptual brackets round everything else. This has obvious benefits in terms of understanding – by bracketing-off the interactions between aspects, we can get a deeper understanding of each one in particular.

The understanding which results from this bracketing process, however, has a rather abstract, artificial quality. It is an understanding of only one particular aspect of any design situation: an understanding whose reliability relies on holding everything else constant. There is a serious problem in using knowledge of this kind to predict what might happen when we propose practical ways of addressing real-world problems: by and large, in real-life situations, nothing can be held constant. Specialised, expert knowledge can therefore only be used to propose ways of addressing particular aspects of complex real-world problems, and there is no guarantee that any of these partial approaches will have positive practical impacts on any real situation. In many real-life situations, the impacts are negative indeed – as we shall see if we scan the next chapter in the Angell Town story.

What we have seen of Angell Town so far has been bad enough, but worse was to follow. The unsurveilled character of Angell Town's pedway decks, together with the fact that they were inaccessible to police vehicles, meant that after a while this complex network turned out to provide a wonderful series of escape routes for criminals – mostly from outside the estate itself – who could easily run in and evade the car-borne police. In an attempt to improve this situation, experts from the police were called in to offer a second helping of professional advice. The police certainly did have in-depth local knowledge of the Angell Town area; but the practical utility this had, from the residents' point of view, was again marred by the specialised, myopic character which is typical of professional knowledge as a whole. Just as the original designers had seen Angell Town as a traffic-safety problem, the police saw it as a crime problem. Once again, any well-rounded understanding of the place as a setting for its residents' everyday lives was obscured by the myopia of specialised expertise, as changes were made to the estate. In an attempt to increase crime-detection rates, pedestrian bridges between the various blocks were removed, to seal off the escape routes which they had previously offered. This change was carried out by local authority design professionals with other experts from the police, and with a minimum level of tenant involvement. No more was sought than an agreement from the tenants' association, which at that stage was little more than a social club for elderly white residents – precisely those who might be expected to be most respectful of professional expertise. Though clearly well-intentioned, the removal of the bridges had unexpected and negative results.

First, the bridges had been used – in ways not intended by the original designers – as play areas; for they formed public spaces which enabled children to meet their peers from other blocks, a little out of earshot from the dwellings themselves. With the bridges gone, children wanting to stay near their homes, or required by their parents to do so, had only the pedestrian ways themselves to play on. Now the noise they generated was right outside other people's flats, and in some locations over the tops of bedrooms. Unsurprisingly, this led to all sorts of social conflicts and accusations of 'children from hell'. Second, not all the blocks had refuse chutes, and with the bridges gone, people found that they had to make sizeable detours if they were properly to dispose of their rubbish, say on the way to work in the morning. Inevitably, some people didn't bother, refuse began to accumulate, other people began to see little point in taking the trouble to be tidy themselves, and local authority management problems compounded the situation. Not only did this lead to an ever-worsening image of neglect and decline, but cockroaches and rats also became an endemic part of the Angell Town milieu. Third, the loss of the bridges meant that people from some of the blocks had to walk through dark and unsurveilled garage areas to get into and out of the estate, unless they were willing to make lengthy detours. This still further raised the level of apprehension and fear on the part of many residents.

All in all, then, the results of the bridge demolition were perceived by the tenants as overwhelmingly negative. Of course, it is again tempting to see all this as the result of stupidity on the part of the particular experts concerned, but that would be a facile explanation indeed, which could only be suggested through hindsight. The point is that the problems associated with the bridge-demolitions could not have been figured out in advance, from the standpoint of abstract global knowledge because the global experts at Angell Town were faced with a doubly unfamiliar situation: neither the innovative nature of the physical form of the place itself, nor the particular local cultural situation came within the purview of their established expertise. And the point is that this kind of unfamiliarity – though often to a lesser degree – is inherent in every situation where specialised, myopic professional expertise alone is brought to bear on particular local situations.

The unavoidable implication of all this is that professional expertise suffers from fundamental contradictions in its own terms, because of the myopic neophilia which is its current hallmark. In theory, it is claimed to lead to 'better' results, and belief in these claims is important in enabling experts to make sense of their working lives. In practice, however, it is hard to see the results of the Angell Town bridge demolitions as anything other than catastrophic in any rational terms: in its present form, professional expertise does not make sense.

To repeat, however: this does not mean that the concept of professional expertise should be abandoned in favour of some swingeing programme of de-professionalisation. Professional expertise brings benefits and pleasures as well as problems, so we are faced with the practical problem of reconstructing our concept of expertise in such a way as to overcome its problems without losing its advantages in the process. What we need, it seems, is a new form of expertise, in which specialised innovation can be focused where it will achieve the most practical benefit for the least cost in terms of unpredictable side effects.

Whether they are myopic and neophiliac, or whether they take on some more closely-focused form, both innovation and specialisation are only made possible through the existence of typological structures. Innovation depends on types because it is types which offer the basic cultural raw material which enables innovation to take place at all: nothing is constructed out of thin air, as we have seen. To the extent that it yields practical benefits rather than dissolving the development team into pure anarchy, specialisation too depends on types: it is only the regime of types in good currency which offers any prospect of focusing and co-ordinating the differently orientated actions of the various actors involved in the form-production process. If we want to develop a new, more closely focused approach to specialised innovation, therefore, we have to consider the relationship between expertise and the typological regime which both enables and constrains its operation.

As we saw in Chapter 3, any typological regime encompasses a complex web of types, spanning institutionalised 'deep' morphological elements such as 'public space', strongly connected to powerful economic, political and legal structures, through to more malleable types such as 'street', 'pavement' and 'paving slab', increasingly near to the cultural 'surface' of the web as a whole. There is an important hierarchical aspect to this progression from 'deep' to 'surface' types, in which each type at a given level sets the preconditions for those at the level next nearest to the surface. No typological regime, however, is constituted solely as a simple hierarchy. Because of the common 'roots' in the 'deep' morphological elements, different typological regimes necessarily share at least some types as common ground between them. Different regimes are distinguished partly by containing different types nearer the surface, but also (and importantly) by different types of relationships between elements at a given 'level' in the web as a whole. In the context of the 'responsive city' regime which we developed in the last chapter, for example, an 'active interface' is a type of non-hierarchical relationship between a building and a public space; whilst a perimeter block is a non-hierarchical relationship between an interface and a public space network – both themselves non-hierarchical relational types. It is these non-hierarchical relational types which, in practice, are central to the distinction between one typological regime and another.

How does all this help us decide how to focus specialised innovation to achieve the best balance between predicted benefits and unpredictable side-effects? There are several implications here, which it may be helpful to express in the form of four simple rules. First, innovation at any particular 'depth' in the web is likely to trigger impacts on all the types 'above' it, for which it sets the preconditions. If many layers are affected by a particular innovation, then given the tacit nature of the knowledge embedded in the various types, the results are likely to be highly unpredictable. The implication here is clear: innovation should be focused as near the cultural 'surface' of the web as possible: the rejection of current types of pavements, for example, is likely to be far more predictable in its outcomes than the rejection of current types of streets as a whole. Second, innovation in regard to the relational types, which bind together several other types at a given level, will not only affect all the types so bound, but also all the elements above these in the web as a whole, effectively dumping all the tacit knowledge embedded in a whole range of types. The level of unpredictability here will necessarily be multiplied many times. The implication here is that we should not reject relational types such as the active interface, or doubly relational types such as the perimeter block unless we find ourselves faced with problems which we cannot find any other way of addressing.

There are situations, of course, where it is not in practice possible to avoid 'deep' or 'relational' innovation. Two further rules, I suggest, help us to get the best balance between innovation and risk here. First, collaboration between the widest practicable range of people, from as many different backgrounds as possible, will help in considering a far wider range of possible unintended consequences than any single individual could cover. And second, these are situations which cry out for post-occupancy evaluation, to see how the unpredictable consequences of deep relational innovation have worked out in practice.

To summarise, then, we need a 'high-resolution' approach to specialised innovation, focusing its effects as near as possible to the surface of the typological regime, and avoiding the rejection of relational types unless this is unavoidable. It is almost always possible to follow this hi-res approach, but when this is impracticable, it is important to involve the widest possible range of expertise in making the design decisions concerned, and to check out the results on the ground afterwards.

To see the potential benefits of the hi-res approach, let us return briefly to the Angell Town Estate. The use of the untried 'pedway' type, rejecting all the tacit knowledge embedded in the active interface type which set the relationship between buildings and public space in the earlier housing which Angell Town replaced, set off a chain reaction of unintended consequences. To overcome these is expensive indeed, requiring major reconstruction of the building stock. This is not to say that the issue of traffic safety which so concerned the Estate's original designers was not a real problem: of course it was. Potentially, however, it might have been addressed through a more focused approach to innovation, as in our own later attempts with the Angell Town Community Project. Here our work was focused as near as possible to the surface of the typological web, avoiding the rejection of proven 'deep' types, and relying on the innovative use of bollards, speed tables and changes of street surface to slow the traffic, thereby adjusting the balance of advantage between pedestrian and vehicular traffic through a 'traffic calming' approach. Not only are these 'surface' innovations inherently more predictable in their effects, but they are also more easily reversible: if the traffic calming turns out not to work in a few years' time, no great waste of resources would be required to rip it out.

None of this should be taken, of course, as implying that the typological repertoire which we developed in the last chapter – or any other typology, for that matter – should merely be accepted as unchangeable dogma. Types have to be used creatively, as raw material to be called on in addressing real-world design problems as they arise in practice. Since these problems will change across time and place, new types will no doubt be created from the raw material offered by those we favour today. The point of the 'hi-res innovation' approach, however, is to concentrate each typological change as near as possible to the surface of the typological web as a whole; a clear example here is indeed the 'traffic-calmed street' type mentioned above, which has progressively moved into the mainstream of European design culture from the 1980s onwards.

Overall, then, specialised innovation is constantly needed as new situations arise, but it has to be regarded rather like a powerful medicine: if it is wrongly prescribed or taken in too high a dose, the patient may die. The golden rule, surely, is the one which tells us 'if it ain't broke, don't try to fix it.' The interactions between users and their urban settings are far too complex to figure out intellectually over the drawing board, except at the most abstract level, so it is only by drawing on the tacit knowledge which has become embedded in loved types, evolved through the practical experience of use over time, that the unintended consequences of any new design can be kept in reasonable check. This is all very clear in principle, but in practice it is not always easy to tell whether particular types are 'broke' or not in relation to particular situations. This is particularly difficult to tell when designers are called on to work in situations which are unfamiliar to them.

Here we come face to face with a further contradiction in the nature of current expertise – this time, a contradiction between two characteristics of professional work. On the one hand, as we saw in Part II, professional expertise necessarily has an abstract, generalised character, because it has to be able to 'travel' between the various different projects on which any expert works. On the other hand, however, this abstract generalised expertise is always in practice applied to particular local situations, which are never 'blank sheets' since they have particular ecological and cultural characteristics which ought to be taken into account if the results of the design process are not to have an accidental quality which contradicts the whole professional ethos. Lacking personal local knowledge, however, the increasingly 'global' professional finds it difficult to take the nuances of any particular situation into account in practice. To bring out the general nature of this problem, which is by no means unique to the design world, let me explore an example which is concerned with scientific experiment rather than with physical design, in a rural rather than an urban situation.

In 1986, UK government scientists were set the task of investigating the effects of radioactive fallout, from the USSR's Chernobyl disaster, on sheep grazing in northern England.[6] In an attempt to investigate how much radioactivity the sheep absorbed through their feeding process, the scientists used their expert theoretical knowledge to carry out a range of practical experiments. The scientists, as Brian Wynne explains,

> ignored the farmers' informed expertise when they devised and conducted field experiments which the farmers knew to be unrealistic. An example was an experiment intended to examine the effects of bentonite spread on affected vegetation in reducing sheep contamination. The experiment involved placing sheep in several adjacent places on similarly contaminated grazing, and spreading different specified amounts of bentonite over each area, then measuring contamination levels in the sheep, before and at intervals after. The farmers immediately observed amongst themselves that these experiments would be useless because hill sheep were unused to being penned up, and would 'waste' in such unreal conditions – that is, they would lose condition and their metabolism be deleteriously affected, thus confounding the experiment.[7]

The upshot of the experts' failure to call on the informal, local expertise of the farmers was that the experiments failed, and were abandoned after a few months (though the farmers' criticisms were never explicitly acknowledged). Clearly, the failure of the experts' experiments constitutes a failure *in the experts' own terms*: the result of their expert work was useless.

It is again tempting to see this as a simple case of stupidity on the part of the scientists concerned, but the roots of the problem run far deeper than that. The scientists' failure stemmed from professional myopia, which in turn had two related causes. First, the scientists' own knowledge was specialised: it was knowledge about nuclear physics rather than hill farming techniques. Second, the scientists placed no value on the farmers' own practical knowledge: to them, its value as expertise was invisible. What could farmers possibly know about the complexity of nuclear physics, after all? These are structural problems, common to most professional disciplines because they are built into the way in which most professional expertise is currently conceived.

In principle, the problems of professional expertise at Angell Town might have been overcome if the experts involved first in designing and then in demolishing the bridges had had the benefit of local knowledge. As Hugo Slim tells us from practical experience,

> Individuals' own accounts of their life and experience usually paint a much fuller picture than most development planners and project workers look for. Above all, personal testimonies connect the various spheres of life, such as family and work, or health and income, which profes-

sionals tend to separate. Relief and development planning is often affected by a kind of inter-sectoral blindness, the myopia of the specialist. The various technical disciplines or professions of development workers mean that they often tackle community development in sectors: health, agriculture, economics, nutrition, law, psychology and so on. People's first-hand accounts of their life and experience tend to flow to and fro between sectors, and to stress the connections rather than the differences. All aspects of a life are intertwined, but it often takes direct communication, rather than a completed survey form, to remind specialists of this fact.[8]

Because the professionals at Angell Town did not collaborate directly with the lay people who were to be affected by the results of their design work, they were necessarily ignorant of the details of how people actually lived their lives within the Angell Town setting. Just like the scientists on the fells, or relief workers afflicted by sectoral blindness, they had no route into the real complexities of the situation in which they were working, and no method except an attempted empathy to try to understand the place from the user's point of view. Having personally met some of the decent and intelligent professionals who had been involved in the bridge-demolition debacle, I am sure that they did try to empathise with the users; but such attempts are always unlikely to succeed in a world where the distribution of resources is stratified in such a way that design professionals are very unlikely to have any personal experience of living on a 'sink' council estate – which is what, at the time, Angell Town had certainly become.

Lacking both collaboration with the residents and any personal experience which might have compensated for direct residents' inputs, it is hardly surprising that the professionals involved ended up defining the estate's problems in their own terms. Again, given the power of initiative which we have already noted, it is no surprise that the experts who played the most telling role in this process of problem-definition were those who were called in first each time: initially the designers, and then the police. The complex roles which the bridges played in the residents' lives were therefore simplified first to a 'traffic problem' and then to a 'crime problem'.

The point here is not that these were wrong definitions: maybe demolishing the bridges, for example, really did make it easier for the police to pursue fleeing criminals. The argument is rather that the complexities of the bridges' roles in people's lives were lost, and more generally that professionals' definitions of problems on the ground must always be abstracted and simplified if they result only from the application of the kind of generalised, 'global' knowledge which professionals need in their travels from site to site. Even in principle, whether on the fells or in the city, global knowledge has to be supplemented by the practical, local knowledge of lay users if problems of this kind are to be minimised.

So far as many design professionals are concerned, this is not a message they want to hear; and it is often disputed by reference to the fact that many past places, including famous set-pieces of urban design which are widely loved today, were designed by the experts of their time, operating in a speculative market without any close links with users. Here, for example, is what Clara Greed has to say:

> particularly when one looks at historical examples of famous urban designs (Bath, for example) relative lack of public participation did not appear to make the design worse – indeed it gave the designer the freedom to get on with the job.[9]

No doubt this is a reasonable picture of what happened in the past, but any implication that an untrammelled 'freedom to get on with the job' will necessarily lead to similarly well-loved results today just does not follow. The undoubted ideological force of this kind of argument is achieved only by ripping the Bath townscape out of its own historical context. There is, after all, precious little similarity between the eighteenth-century develop-ment process, with its technological limitations, small-scale funding pro-cesses and positive relish for using well-tried historical types, as compared with the more recent regard for relatively untried innovations, implemented on a large scale, within a context of ever-expanding technological possibi-lities. For every loved eighteenth-century street or square, we are all surely aware of some huge, failed experiment nearer our own time.

From our perspective, then, there are serious problems inherent in design culture's systematic devaluation of lay knowledge. The drive for innovation and the fragmentation of professional expertise might often together encou-rage designers to set off in directions which are negative from our point of view, but in principle, perhaps, the situation might always be retrieved if users too were seriously involved in the process of design. How might the 'expert' role be re-imagined to encourage this to happen?

Since the idea of 'the expert' cannot just be imagined away, if for no other reason than that it is central to the designer's own economic survival in the professional services market, the only way forward here seems to depend rather paradoxically on an expansion of the idea of expertise – an expansion to make it encompass not only the global expertise of the design profes-sional, but also to include the considerable fund of in-depth knowledge about particular places which users have to build up through everyday practical experience, if they are to manage their lives in effective ways. In other words, users have to be re-imagined as 'local experts', whose partner-ship in the design process offers an otherwise unobtainable source of local knowledge, indispensable to the efficiency of expertise on its own terms.

I am not, at this point at least, arguing that lay knowledge ought to be incorporated into the design process because this will empower users. I do in fact believe this, but at this point I am more concerned to demonstrate that even if one does not value user-empowerment for its own sake, the fact remains that any conception of expertise which persistently devalues lay knowledge is internally contradictory, because it is demonstrably unlikely to produce good results on its own expert terms. Users' responses to designers' proposals can only be helpful, however, to the extent that the users concerned can understand what is proposed. To tap into local expertise, therefore, designers have to develop new skills in explaining the objectives behind what they propose, and the reasons why they think their proposals will achieve these objectives in practice. Unless designers are able to develop 'design rationales' to explain these matters, as an integral aspect of their everyday practice, building explanatory bridges between their own global knowledge and the local expertise of users on the ground, the user–producer gap must continue to yawn; and designers will continue to cut themselves off from the local expertise which they need to perform well even on their own terms.

By this stage in the argument, it should be apparent that there are serious limitations, on its own terms, to the efficacy of current professional expertise. It is important, however, not to throw out the baby with the bathwater. Recognising the limitations of professional expertise, and acknowledging the importance of lay users' knowledge as local expertise, should in no way be understood as diminishing the importance of design professionals, nor of the global knowledge which they use. Whilst it is essential to draw on local expertise if better places are ever to be produced except by accident, it is nonetheless equally important to recognise that lay knowledge by itself is not cast in a form which can make any effective impact on the complex development process of today. Just as global expertise has particular, socially produced characteristics which limit its usefulness in the production of better places, the in-depth knowledge of the local expert has serious limitations too.

Most of the in-depth knowledge about urban places which lay people build up is practical, tacit knowledge, embedded in their everyday practices, and well below the threshold of consciousness. The people of Angell Town, for example, know how the bridges are used, but they do not carry most of this knowledge around in a conscious form. As we already saw, this is a matter of practical necessity: if everyone was all the time conscious of the complex models of their relationships with built form, which they build up in order to live their lives effectively, there would be no room in their heads for anything else.

There are, of course, bound to be some aspects of any user's knowledge about the relationships between built form and social life which are 'remarked on', and thereby stand out, at a conscious level, from the generally tacit taken-for-granted background. In practice, this 'remarked on' knowledge is usually focused primarily on problems rather than on potential solutions. Given the complex division of labour in the modern development process, this is hardly surprising: users discover the problems places have, through their lived experience; but this experience provides them with few opportunities to develop similar discoveries in terms of putting things right. Making changes to built form is the sphere of the design professional, after all. The upshot, however, is that local expertise is all too often couched only in terms of what is wrong with recent transformations – terms whose negative character makes professionals feel justified in maintaining the gap between users and themselves: 'There's no point trying to involve the local people. All they do is moan about everything, and when you ask them what they *do* want, of course they can't tell you.' Bringing global and local expertise together in a fruitful union is therefore no easy matter: it requires the development of new skills to overcome the problems outlined above.

Skills of this kind could themselves form part of the global expertise of the design professional, since their use would be independent of particular locations. What design professionals need here is the ability to tap into the tacit, problem-orientated knowledge of lay users, and convert it into the kind of verbalised, solution-orientated knowledge which can be used in the design of new urban places. A further stage in the Angell Town saga may help to illustrate how this conversion from local to global expertise might be managed.

Our own involvement in developing an urban design strategy for addressing the estate's design problems began with a clean sheet of paper, rather than with us making proposals to be discussed. This meant that there was no structured agenda of design issues to discuss – rather residents were asked to talk first about their experience of problems on the estate. Not surprisingly these were not issues whose links with urban design were immediately obvious, so our role was to help uncover whether such links existed, and if so what they were. An example may help to make this clear. At the first meeting, we were disconcerted to discover that dog shit was the topic which people were most anxious to discuss. The links between this prevalent nuisance and urban design were initially unclear but were uncovered as the meeting progressed, through a process of give and take between the two sets of experts. 'Why is there so much dog shit, anyway?' we asked. 'Because there are so many dogs', came the reply, delivered in a tone of voice which made it clear that the speaker found this a very stupid question. 'But why are there so many dogs then?' At this point, the answers began to uncover a dog-culture which was utterly new to us middle-class urban design consultants, to whom 'dog' had hitherto been equivalent to 'family pet'. It became clear that to many dog-owners at Angell Town, the dog was at least partly a cost-effective, high tooth-powered security device; and someone recited what was later to become a familiar *bon mot* about neighbouring Stockwell Park: 'The estate where even the Rottweilers go round in pairs'.

> 'But why do people feel so threatened anyway?'
> 'Because the place is always deserted.'
> 'Most of the time nobody can see what's going on when you're outdoors . . .'

By now, we were clearly into urban design issues, which the consultants felt able to grasp. And the point is that these were issues about community safety, the perception of threat in public space and so forth, which our previous design thinking had only engaged in the most peripheral way. These were issues which had almost entirely been ignored in our own frame of reference hitherto, and yet they were massively top of the residents' own agendas. Our professional frame of reference itself was thereby subjected to a powerful critique and was subsequently enriched, much to the benefit of our design approach in future work. It is hard to see how that could have happened without engaging with this previously unfamiliar culture in a process of action research. And it is hard to see how that could have happened without the residents driving the design process, according to their own agenda, from the outset.

This is not to say, of course, that we did not incorporate our own agenda of preferred types into the process – merely that we tried hard not to introduce our own pet concerns until a basic framework of issues had been articulated by the residents, and we did our best to avoid manipulating the situation to make it come out how we wanted. In the nature of things, we probably failed in this to some extent – it is difficult to become a 'neutral instrument' after all – but the fact that we are still around, still with a positive relationship with the Angell Town Community Project, nearly a decade later, suggests that the residents feel they are getting enough of what they want. Interestingly enough, the scheme has also been well received in mainstream architectural circles. It was thought worthy of a major ten-page Building Study in Britain's *Architects' Journal*,[10] while *Resource for Urban Design Information* tells us that 'the scheme was successful in combining resident consultation and good urban design practice.'[11]

Through processes of the kind employed at Angell Town, users and professionals effectively become collaborators in an ongoing programme of action research, integral to and therefore relevant to the process of design, and focused on the relationship between built form and real-world social issues. Through this programme, tacit local knowledge is continually transformed into more abstract, verbalised professional expertise, which then forms part of a renewed professional culture which can be transported to other projects. In my own experience, attempts to develop such techniques of collaborative working between local and global experts in places like Angell Town have indeed changed the professional culture of urban design over the last decade or so. The collaborative process of design-orientated action research – sometimes called 'reflective practice' – generates professional knowledge which seems to be far more directly relevant to real-world design processes than most of what comes out of more traditional academic research. It is therefore readily absorbed into mainstream design culture, when promoted through the professional media: our own writings and professional development seminars, for example, have led directly into a wide range of local government programmes of design guidance.

This process of converting tacit lay knowledge into professional expertise should not, however, be seen as generating information in the form of fixed, universally applicable products. The professional expertise which results should still be thought of as constituting an agenda for design, whose relevance has to be tested anew in each local situation. The key point about local situations, after all, is that they differ from one another, so that even local people who share the concerns we developed in Part III, and value qualities such as vitality and local identity, might have their own particular local conceptions about where the best trade-off between them might lie – a point which comes over very clearly in this account of debates around the production of a *Plan d'Occupation des Sols* for the French municipality of Asnières sur Oise, by my colleague Ivor Samuels:

> In these debates it was sometimes difficult to abandon cultural preconceptions. For example, the long blank walls enclosing the gardens to the large mansions do not encourage street vitality and regulations were considered to reduce their length – until it was pointed out that it was a characteristic of the village and indeed of many villages in the Île de France, which was acceptable to the population and should be maintained.[12]

Overall, then, our exploration of the contradictions which are inherent in current notions of 'the expert' has generated a range of useful insights about how professional cultures should be reconstructed to avoid the problems which the myopic neophilia of much current work places in the way of competent expert performance. First, we explored the internal contradictions which flow from neophiliac attempts to reinvent the wheel. We came to see that there should be no bar to calling on the web of user-positive types which we identified in the last chapter, merely because elements of that web have been around a long time. We saw, too, the need for designers and users alike to call openly on this repertoire, so as to be aware of the larger implications which even the smallest design intervention might have, and so that when we innovate, we do so in a high-resolution mode, as near as possible to the surface of the typological web as a whole. Finally, we came to understand that tacit lay knowledge constitutes an indispensable base of local expertise, which should be drawn on directly wherever possible, and which should be taken as the touchstone against which professional, global expertise has continuously to be monitored and developed.

To the extent that these ideas are valid, then any support we can give in moving them into the mainstream should help to create a cultural climate within which design professionals can more easily contribute to making better places – better according to canons of professional expertise and users' values alike. No matter how intellectually convincing the arguments we have explored in this chapter, however, this reconstructed expertise alone will never form a sufficient basis for the passionate crusade which will be needed to push through the production of better-loved places, taking up the opportunities presented by ecological fears in the face of pervasive opposition from short-term economic pressures. Expertise alone, in whatever form, has to be supplemented by emotional commitment. Within current design culture, it is perhaps the art dimension which offers the most fruitful potential for developing this commitment in practice. But how might we relate this chapter's ideas to *any* conception of art? To anyone who makes the traditional links between art, individual genius, innovation and originality, it must surely seem that what we are proposing here runs counter to every established artistic value. How *can* it be art? To engage the support of the artist as well as the expert, that is the question which the next chapter must address.

12 Artists in a common cause

We have seen that the paradigm of 'the artist' plays a key role in helping both designers and users make sense of what designers do. We have also come to understand that their 'artist' status offers architects, in particular, their unique selling point in the capitalist design-service market. To abandon this claim would be economic suicide, so the conception of designers' work as art seems set to stay. This presents both problems and opportunities in terms of the user-positive working practices we seek.

On the positive side, the 'art' paradigm offers the potential for a way of working which can give aesthetic–sensual support to supplement the cold rationality of expertise. In particular, advantages flow from the fact that the art tradition gives a positive value to intuition as a valid mode of accessing knowledge: in principle at least, it places no bar, as expertise does, to calling on tacit, practical knowledge, such as that embedded in types, which cannot be rationally supported. In addition, important elements of the recent art tradition share a rebellious desire to reject the mainstream culture which supports the status quo. Unlike much of the expertise which we explored in the last chapter, art has *attitude*. All this is very positive from the point of view of art's potential in making better-loved places, but we saw in Chapter 6 that the current mainstream of the art tradition also has many shortcomings from our point of view. Its conception of the creative process is often narrowly framed in terms which exclude users and devalue the contribution of all but male artists. Its conception of the art work itself focuses narrowly on the hardware of urban form, and then only in visual terms, encompassing only three dimensions. Its evaluative criteria are shot through with mystique and shared only by those 'in the know', and in practice have evolved mostly to support the reproduction of profitable forms. As presently conceived, the practical problems of the art paradigm appear to outweigh by far its positive potential, from our point of view.

Similar points could be made about the negative impact of current conceptions of art in many areas of human life; this is a problem with a far broader focus than urban form alone. It is perhaps not surprising, then, that criticisms like these have led many theorists of culture to reject the idea that art can play any positive role whatever in human affairs. There is a widely broadcast voice, according to Fred Inglis, urging us to 'stuff art; it is another name for effective and affective power.'[1] If it really is a source of effective power, however, it should surely not merely be ignored. On the contrary, any available source of effective power should be taken up and used to work towards the production of better, more user-positive settings. If those who want better settings merely ignore the potential power of art, then in practice they court the danger that it will end up being used against them to support the status quo, as the US writer bell hooks suggests:

> Recently, at the end of a lecture on art and aesthetics at the Institute of American Indian Arts . . . I was asked whether I thought art mattered, if it really made a difference in our lives. From my own experience, I would testify to the transformative power of art. I asked my audience to consider why in so many instances of global imperialist conquest by the west, art has been other [sic] appropriated or destroyed . . . It occurred to me that if one could make a people lose touch with their capacity to create, lose sight of their will and their power to make art, then the work of subjugation, of colonization, is complete.[2]

These dangers seem very real today. It is becoming increasingly obvious to many artists themselves that art can never float free in some depoliticised realm of the aesthetic, since it constitutes a potentially powerful force for change which in practice is all too often appropriated by powerful interests to serve their own ends. Hans Haacke, for example, refers to what he has learned from the attempts at art censorship which have been made by US Senator Jesse Helms, 'an important figure in the . . . extreme right', who 'hates women and gay people who have the audacity to claim their legitimate rights'.[3]

> I believe that Senator Jesse Helms taught artists, and other people who care about free expression, an important lesson. He reminded us that art products are more than merchandise and a means to fame, as we thought in the 1980s. They represent symbolic power, power that can be put to the service of domination or emancipation, and this has ideological implications with repercussions in our everyday lives.[4]

From our perspective, the most central of these implications is the imperative to seek ways of using the power of art to emancipate the user–designer alliance, rather than to increase the weaponry of domination available to the power bloc in its efforts to keep the status quo in place. There is a strong tradition of emancipatory effort within the world of art, and a long roll of honour encompassing artists who have sought ways of articulating this tradition through their work, as Fred Inglis points out:

258 *Artists in a common cause*

Solzhenitsyn or, say, Kozintsev, Simone de Beauvoir and Isabel Allende; or, in a third realm of recent witness-bearers against power, Ralph Ellison, Billy Holliday and James Baldwin, were all *artists*. They faced power with a power as mysterious, familiar and creative as a radioactive field. They defied the casual horribleness of old coercion.[5]

The search for ways of using the power of art to push for emancipatory change has also had a powerful impact on making better-loved places, by inspiring important projects of community action. Edinburgh City Councillor David Brown, speaking of the city's Craigmillar Festival Society, tells of 'the creative power of art' as the prime mover in the Society's very effective efforts to improve the huge Craigmillar housing estate.[6] Dora Boatemah too tells of the power of art, articulated through the biennial Angell Town Festival, when she speaks about the importance of its 'power in getting people together' in promoting the user-positive changes there.[7]

If we want to put this power at the service of making better-loved places – if we want to tap into aesthetic–sensual resources which are beyond the purview of the often cold, rationalistic world of 'the expert' – then we must not reject the art paradigm out of hand. The problem in our terms, therefore, is not that design professionals often think of themselves as artists, nor that many everyday users encourage them in this belief. The problem lies rather in the kinds of artists they feel themselves to be. How then, should users and designers reposition themselves within the overall field of art, to gain sensual and aesthetic support for working towards better-loved places, to reinforce the type of rational expertise which we developed in the last chapter?

By any reckoning, the field of art is a broad church containing many movements and tendencies, espousing many different systems of values. To reinforce our attempts to overcome the power bloc's pervasive support for current mainstream types, we need an approach to art which takes a combative and challenging approach towards the status quo. Within the art fields of the capitalist era, the greatest potential in these terms is offered by those movements which form part of the avant-garde tradition. Occupying a central position within the overall field of art in recent times, the avant-garde defines itself in relation to a metaphor drawn from military life, implying a view of artists as forming an 'advance guard' – cultural 'shock troops', so to speak – taking a challenging and combative stance in relation to conventional ways of making form.

In principle, this is a highly attractive stance from the point of view of making better-loved settings, precisely because it takes a challenging approach to standard ways of doing things. Not all such challenges are necessarily positive from our perspective, however; and in practice there is an important distinction to make between different senses of the 'advances' which the advance guard might make. Within the avant-garde tradition there are two such senses, between which it is important to distinguish.

On the one hand, there is one movement within the avant-garde which conceives of itself as advancing into unknown territory. From this perspective, it is the process of exploring new artistic areas itself which matters, rather than what is discovered or created there. In practice, this amounts to a delight in originality and innovation as positive for their own sakes, requiring no reference to 'outside' indicators of quality. Artist-designers in this wing of the avant-garde, driven by the search for originality, regard all established types negatively, as constraints on their personal creativity. As we can see from our structurationist perspective, they can never escape altogether from the typological repertoire in good currency, so their rebellion has in practice to proceed by using current types in unexpected ways – reversing or inverting them, as we saw in Chapter 3, for example, with the work of Le Corbusier. This destroys the tacit knowledge embedded in these types, through the same neophiliac approach whose many practical problems, from the user's point of view, we explored in the last chapter. Whether we call neophilia 'art' or 'expertise' makes no practical difference to the users who are faced with its likely negative impacts.

This does not mean, however, that the avant-garde tradition as a whole can offer us nothing. It has another wing, whose members wish only to overthrow types which support the current power bloc's status quo, rather than vainly seeking to escape from existing typological structures altogether. Again, there are several versions of this general approach.

At the most obvious level, we find art as political propaganda – art for 'our' cause. This has certainly given rise to some powerful political statements, from the agitprop of the Russian Revolution to the street graffiti of the Palestinian Intifada. In the end, however, all such overt messages are likely to be self-defeating in a developed capitalist context, for the simple reason that capitalist institutions have turned out to possess an extraordinary capacity to absorb the critical aspects of avant-garde art, as currently constituted, and to appropriate them for their own purposes – those of capital accumulation and of reproducing the system which makes capital accumulation possible. George Weissman, speaking on behalf of the Philip Morris cigarette company about sponsorship for artists, makes crystal clear the reason for forging these links between art and the capital accumulation process: 'Let's be clear about one thing. Our fundamental interest in the arts is self-interest. There are immediate and pragmatic benefits to be derived as business entities.'[8]

As Bernard Tschumi reminds us, 'Duchamp's urinal, after all, is now a revered museum artifact; revolutionary slogans of 1968 Paris walls gave new life to the rhetoric of commercial advertising.'[9] So Picasso's paintings hang on the walls of banks to show how cultured bankers are, whilst photographs of misery and suffering, no doubt originally taken to stir the conscience of the viewer, are used by Benetton to sell fashionable clothes. And, as we have seen in some detail, the avant-garde of the arts of urban place has contributed both its forms and its ideas to the capitalist mill. Viewed from outside, even the 'political' wing of the avant-garde seems like nothing more than a struggle for prestige and cultural capital, within an introverted elite framework. In practice, art has been co-opted into supporting the status quo. The shock troops of the avant-garde have ended up sleeping with the enemy, thereby failing on their own terms.

The process of co-optation which underlies this failure operates at two related levels. Both of them are difficult to avoid. First, art is a great source of new, saleable commodities. Critical propositions, aimed at changing the status quo, can have no practical impact unless they somehow strike a sympathetic chord amongst a constituency wide enough to bring about practical change. The moment this constituency is formed, however, its size offers the potential for profit, since all its members are also consumers. In the constant search for new opportunities for capital accumulation in the competitive marketplace, therefore, entrepreneurs become motivated to try to take over the radical message and incorporate it into commodities for sale. Metaphorically speaking, the revolutionary slogan gets printed on a tee-shirt for market exchange. Converted into radical chic, it thereby not only makes a profit for the entrepreneur, but also becomes trivialised in the minds of many of those whom it originally moved.

A very clear example of this level of the incorporation process can be seen in the fate which befell the politically radical rock music of US West Coast counterculture during the 1960s. As John Storey sees it, this music had originated as an aspect of political protest:

> For the political folk singers music had been a means of class mobiliza-tion, of organization, the muse of solidarity. For the counterculture it was the central and unique mode of political and cultural expression.[10]

Soon, however, the size of its audience became seen as a source of profit as much as threat – a source of profits which could be used to help in funding enterprises such as the US involvement in the Vietnam war, to which the music itself was totally opposed; as Keith Richard of the Rolling Stones discovered:

> We found out, and it wasn't for years that we did, that all the bread we made for Decca was going into making black boxes that go into American Air Force bombers to bomb fucking North Vietnam. They took the bread we made for them and put it into the radar section of their business. When we found that out, it blew our minds. That was it. Goddam, you find out you've helped kill God knows how many thousands of people without even knowing it.[11]

Exactly the same drive towards the commodification of protest also affects the arts of urban form, in ways which are hard to avoid in practice. Working on a millennium project for Bethlehem – a town which needs all the practical economic support it can get – we even caught ourselves wondering whether the graffiti of the Palestinian Intifada might not prove a valuable tourist asset. At a second level, the arts of urban form are incorporated into the capital accumulation process through the market-driven process of replicating and typifying those innovations which make urban places more easy to control, in the interests of keeping the status quo in place. In Part II, we made a detailed exploration of how this process works.

Does all this mean that even the most resolutely political avant-garde is doomed to failure in its attempts to subvert the status quo? What avenues of change are possible for those who want to reconstitute a truly critical avant-garde for today? How might the pervasive and cumulative pressures towards both these levels of co-optation be avoided?

It seems to me that the only avenue of escape from co-optation through commodification, at least in the case of art works as physical artifacts, lies through creating the work in some physical medium which cannot itself be bought and sold in the marketplace. As I have already argued, publicly owned space is the only physical medium which cannot be commodified so far as urban settings are concerned; so art which takes public space as its physical medium might, in principle at least, be able to avoid the commodification process.

The qualities of the sensual, aesthetic experience of public space – a level of experience highly relevant to public space as an art medium – are, however, very much related to the design of the buildings and other artifacts which give it sensory definition; and these themselves are typically produced as commodities, for profit, through the medium of private-sector funding. If public space is not to lose the positive character of its autonomy from the capital accumulation process, becoming reduced merely to the space left over between commodified buildings, then the buildings themselves will have to be designed to contribute to the formation of public spaces, themselves treated as autonomous art works. Since it is only these aspects of design – in so far as they can be achieved at all – which can create the potential for an avant-garde art of public space, it follows that it is they, rather than the building itself, which constitute the art work.

Viewed from this perspective, then, any attempt to produce an avant-garde art of built form must involve the re-imagining of buildings in terms of their impact on the formation of public space, rather than as discrete 'things' in themselves. It is this incremental contribution which each particular building project makes to the user's experience of public space which is important from the avant-garde perspective; what matters here, then, are the characteristics of the interface between any particular building and the adjoining public space, and the degree to which the privacy gradient inside it promotes a positive contribution to the public realm. To engage with these issues in the process of design, as users experience them, artist-designers will have to create new ways of imagining this experience and of representing it to themselves and communicating it to others – ways which will have to be very different from the small-scale drawings and models, divorced from their surroundings and often artificially representing impossibly distant and abstract viewpoints, which are typical at present.

The second level of co-optation – incorporation through collusion in propping up the status quo – can only be overcome through public spaces which offer support to the open, exploratory aspect of its users' desires, which can themselves never be totally controlled. No matter how effectively desire is channelled down from its open, exploratory nature into commodity-centred consumer choice, there is always something more – something potentially rebellious, which gives popular culture, for example, its critical edge. Any art practice which is intent on avoiding incorporation as an integral part of the capital accumulation process, therefore, will only be able to do so (if at all) by working towards settings which support their users' desires for non-commodified choice. As we saw in Chapter 7, the qualities of variety, accessibility, legibility, robustness, identity, cleanliness, biotic support and aesthetic richness have a particular importance here. To the extent that these qualities can be supported through the art of urban design, to offer ordinary users enhanced resources for their own life-projects, they represent tactical defeats for those who support the status quo. As Terry Eagleton points out, 'the only thing the bourgeoisie cannot incorporate is its own political defeat';[12] so here lies the key to an avant-garde practice which does not have to end up sleeping with the enemy.

To the extent that this analysis is valid, it is clear that a genuinely avant-garde approach would be very different to the reactionary arts of urban form which pretend to the avant-garde mantle today. A true avant-garde would have to fulfil two main conditions. First, in order to work with the medium of public space as a whole rather than focusing merely on separate object-buildings and other artifacts, it would have to be able to bridge across the arid division of labour which is typical of today's 'development team'. Second, in order to remain open to the exploratory, rebellious desires of popular culture, it would have to bridge the yawning producer–consumer gap which characterises the current situation; and in bridging that gap it would have to pay particular attention to those users who, because they are most disadvantaged in terms of gender, age, ethnicity and disposable income, are most likely to be located at popular culture's critical edge.

This bridging role can only be played by some kind of shared typological repertoire. To maintain its avant-garde potential, the art of urban design has to embody a repertoire which offers users a high level of non-commodified choice of the kind of resources they desire for constructing their own life-projects. The repertoire of deformed grids, complex use-patterns, robust plot developments, positive privacy gradients, perimeter blocks and native biotic networks, which we identified in Chapter 10, offers the richest potential we have so far found in this regard. We have already seen, however, that the overt use of any typological repertoire, within a Western fine art tradition, is likely to be interpreted as 'unoriginal', and therefore negative, by the cognoscenti. This raises serious practical problems, because any viable avant-garde would need as much cultural capital as it could muster, to promote its subversive projects in the face of opposition from those who support the status quo. How might the arts of urban design be re-imagined to overcome this problem?

The greatest potential here, I think, is offered by a particular vision of art, in relation to built form and the public realm, which by implication runs from Pythagoras's musical analogies through Friedrich von Schelling's Romantic eighteenth-century image of architecture as frozen music[13] through to recent developments in what has come to be known as 'new genre public art',[14] in which, for example, the products of SPARC's *Cultural Explainers* project in Los Angeles, though they are physical artifacts, are considered by their creators to be 'art as public performance'.[15]

Viewed from this perspective, we can begin to see the creation of urban design works as performances of the 'responsive city' repertoire, which constitutes, at least potentially, a 'score' for co-creation. From the point of view of the avant-garde artist, then, we come to the same practical conclusion as that which we reached from the expert's perspective. Any avant-garde which is to avoid absorption into the capital accumulation process, like any expertise which is not to fail on its own terms, can only be founded on a repertoire of types whose creative performance can offer the maximum support for the free play of its users' own open, exploratory desires, rather than constraining them in the power bloc's interests. It is therefore only types with this potential, rather than desocialised, abstract 'space', which can form the medium of an avant-garde art of urban design. Viewed in this light, a repertoire of types such as the one we developed in Chapter 10 does not represent a constraint on artistic creativity, but rather constitutes its fundamental ground; rather like an open choreographic structure, which both inspires the dancer and, in the process, distinguishes the art of dance from mere random hopping about.

Of course there are many ways in which urban design is not like other performance arts such as dance or music, just as there are many differences between architecture and fine arts such as painting or sculpture (someone – Suzanne Langer, I think – once rather memorably pointed out that music is not melted architecture). By foregrounding the common factors which enable urban design to be imagined as a performance art, however, we might be helped to take a creative leap which allows us to think and feel about design in new ways; which in turn might help us break free from the cultural bonds in which we are imprisoned by the dead hand of the conventional fine art approach.

First the re-imagining of urban design as a performance art foregrounds, in a very direct way, the idea of the collaborative creative act. Here the co-creative performance, to some extent at least, transcends not only the barriers between one artist and another, but also that still more crippling gap between production and consumption, between the artists and the audience. Viewed in this light, urban design can be seen as an art involving two levels of performance: first in producing given settings themselves, and then through the productive use of them through the exploratory aspect of desire which drives the arts of everyday life.

The co-creative working of artists in the world of performance is cele-brated – as it is usually not, for example, in the world of the design professions – through the everyday vocabulary of the performance world, in which 'groups', 'bands' and 'companies' abound; and in which producers, directors, artists and technicians of all kinds are given proper title credits for their involvement. In recent music, for example, one compositional philo-sophy, under the leadership of the composer John Cage, is seen as note-worthy by the musicologists Grout and Palisca for 'allowing performers to take a greater role in determining the final form and even the pitches and rhythms of a work'.[16] Re-imagining design as performance, then, helps us towards a more positive sense of the currently unloved 'development team', and might thereby help us take a first step in breaking down the arid division of labour which currently holds sway within the capitalist devel-opment process.

The same 'performance' perspective also helps us towards a more positive evaluation of the possible roles of all sorts of people who are currently undervalued, at least by implication, within the design professional's world. We have already seen, for example, the negative effects of systematically denying 'genius' status to women – an exclusion which we might be helped to remedy when we recall the central role of women as performers, for example, in music, theatre and film. By the same token, recognising the obvious contribution made by African, Asian and Latino cultures in the performance world of music, for example, may help us take a more creative approach in searching for ways in which multiple cultural values can together find expression in a truly public art of urban form.

The re-visioning of design as an art of public performance, rather than as a fine art based on more inner-directed studio work, might also be expected to affect the kinds of people who are attracted to join the world of profes-sional design in the first place. The current studio-centred fine art para-digm, as we saw in Chapter 6, has particular attractions for those of an introverted disposition. The re-imagining of design as performance, with the new connotations of extroversion which that would involve, might encourage more extroverts to join – a shift which, in turn, might be expected to develop a culture in which it seems more natural to reach out both to other design professionals, and to users too, across the fractured division of labour.

Re-imagining urban design as performance, indeed, suggests a range of other ways of thinking and feeling our way beyond the current crippling gap between production and consumption, between the artist and the audience. Here too we might forge a creative analogy with the world of music; drawing, for example, on Christian Asplund's comments about the music of the avant-garde composer Frederic Rzewski:

> The people in the space are divided into two groups but are not called 'performers' and 'audience'. They are rather 'musicians' and 'non-musicians'. The non-musicians play a simpler and freer version of the piece: any sound in steady eighth notes . . . The relationships of collectivity, trust and mutual improvement that Rzewski facilitates between performers, audience and composer are the most profound aspects of his work.[17]

Audience participation, in the musical context, is by no means confined to the 'difficult' music of composers like Rzewski, nor to the work of large bands, orchestras or groups. Even solo performers, for example, know that they have audiences to relate to; and they know too that the quality of that relationship – the degree to which the performer is able to involve the audience in the creative act – is often central to the quality of the performance itself. David Rowe, for example, points out the importance of the audience as active participants in the world of rock music, in which 'through various dispositions of the body, stars and fans alike "materialize" and articulate rock culture'.[18]

In the field of theatre, too, it is widely acknowledged that the audience's productive capacities make important contributions to the quality of the performance; one strand of avant-garde theatre has for many years tried to capitalise on this by developing techniques for audience participation. As far back as 1930, for example, *The Measures Taken* was put on by the German playwright Bertold Brecht with three large workers' choirs, constituting both the audience and the 'control chorus' which formed a central part of Hanns Eisler's musical score. Questionnaires were distributed to gauge audience reactions after the first performance, and the critical debate which followed led Brecht to rewrite a crucial scene with a completely new meaning.[19]

The fundamental importance of the co-creative link between artists and audience is celebrated in the everyday vocabulary of performance art: 'the concert' – a word which itself implies something 'concerted' – stands for an event which encompasses both artists and performers together. Much emphasised in the development of avant-garde theatre, through ideas such as 'theatre in the round' or the 'thrust stage' which brings performers into the centre of the audience, this re-imagining of the artist and the audience as together constructing the performance opens up a conceptual space for re-imagining the urban design process not as something which ends when the cheque is handed over at the completion of a particular building contract, but as something which continues through the activities of users, in a seamless creative process in which the physical structuring of space is merely one particular moment, which 'sets the scene' for users' performances to come.[20]

The concept of the audience as co-creative participant in the art work through dance, called to mind for example by the rock music analogy, has also been related more directly to urban design through a focus on the 'ballet of the streets', in Jane Jacobs's memorable phrase.[21] The feeling for public space as a setting for the dance of everyday life, perhaps explored most intensively by the dancer and choreographer Anna Halprin and her partner, the landscape designer Lawrence Halprin, and codified in their 'motation' approach to urban space as an open choreographic score,[22] also underlies much classic thinking in the world of film. For example, the director Sergei Eisenstein – himself originally trained as an architect – saw the Acropolis of Athens as 'the perfect example of one of the most ancient films',[23] and expressed his feelings about it like this:

> At the basis of the composition of its ensemble, at the basis of the harmony of its conglomerating masses . . . lies that same 'dance' that is also at the basis of the creation of music, painting and cinematic montage.[24]

Eisenstein himself is well known for the highly suggestive analogies he drew between the perceptions of cinematic and urban space, but the medium of film has much to offer at other levels, in terms of helping to think ourselves beyond the constraints of 'fine art' orthodoxy. For one thing, cinema is the co-creative art par excellence, at least at the artefactual level: look at the credits at the end of any film, and you will see the extent to which this is acknowledged. Just as interesting, from our point of view, is the way in which film offers different audiences – or different sectors within the same audience – different aspects to value. The French social anthropologist Pierre Bourdieu, for instance, points to the fact that different groups of people value actors over directors, and vice versa, depending on differences in educational capital.[25]

Analogies between film and the arts of urban form are, of course, extremely fashionable nowadays, and we have to be properly aware of their dangers as well as what they can offer us in a positive sense. The critic Kester Rattenbury offers us a salutary warning here:

> Recently, comparisons of architecture to film has [sic] become one of the highest forms of critical praise. It is a rich analogy which has generated wonderful ideas and deepened some forms of architectural criticism. And it is profoundly mistaken. Let's get this straight. Architecture is essentially, inherently different from film.[26]

This is so obviously true of any literal analogy between film and architecture – or film and any other aspect of urban form, come to that – that one wonders why the point is worth making – until one remembers Bernard Tschumi pointing to the different interpretations of postmodernism in art as compared with architecture: 'postmodernism in architecture became associated with an identifiable style, while in art it meant a critical practice.'[27] I hope it is clear that I am not here interested in the stylistic level: to say it again, I am not proposing that urban design is literally like film; rather I am suggesting that thinking about urban design *as if* it were a kind of co-creative performance, as film partly is, might help us to break free of current orthodoxy's bonds, to begin to develop the critical practice whose absence Tschumi regrets.

The potential for communal creativity which is opened up by the concept of performance art does not, of course, preclude the role of the individual genius which is the more traditional model in the spatial arts. In practice, this is an extremely important issue, because it relates to questions of the 'authorship' of the art work, which are central to the economic survival of the artist in urban design's commercial marketplace. The artist Judith Baca, co-creator of *The Great Wall of Los Angeles*, points up the issues involved in this kind of artistic production when she asks 'whose art is it, the kids, the homeless or yours?'[28]

From her own experience, Baca sketches out the field of possibilities here, in an undefensive, open way which is potentially very helpful for thinking through this dilemma in urban design:

> In some productions where you are going for the power of the image, you can get a large amount of input from the community before the actual making of the image, then you take control of the aesthetic. That's one model. Another is a fully collaborative process in which you give the voice to the community and they make the image. Both of these processes are completely valid, but there's very little room for the second because artists take such huge risks becoming associated with a process that might not end up as a beautiful object.[29]

Given the importance of the aesthetic dimension to those design professionals who see themselves as artists, which we explored in Chapter 6, it seems to me that most urban design productions are adopting the low-risk strategy here, 'going for the power of the image' in Baca's terms; so her first model seems the one which is most likely to prove culturally acceptable both to urban design professionals and to those who hire them in the market for design services. In terms of this first model, it is clear that performance art still leaves space for the individual artist to shine: the soloist in the orchestra, the prima ballerina, the star actress or actor, the diva or even the 'name' director can all be involved. Set within the socialised context of the performance, rather than conceptualised as creating art *ex nihilo*, and set free from the deadening process of stereotyping around ethnicity and gender which suffuses analogies with painting and sculpture, the individual artist can add lustre to the communal work in a way which is entirely positive from our point of view.

Taken together, all these examples clearly show that a concern for max-imising the involvement of lay people in the design process need not be taken as denying the importance of avant-garde leadership in the produc-tion of better-loved urban places. On the contrary: viewed through the other end of the telescope it is only the involvement of lay people, making the design their own through communal creativity, which can build the poli-tical support needed for achieving an avant-garde art of public space in the teeth of the speculative development process. Seen from this viewpoint, leadership implies the initiation of performances in which others are willing to take part. The mere imposition of the artist's will on users who are themselves excluded from the creative process, in the traditional model of the failed, co-opted avant-garde, is not the mark of leadership. It is merely a crude example of the artist as bully.

The conception of urban design as performance art, then, offers a route towards the imaginative reconstruction of the role of the urban designer as artist, to get beyond the internal contradictions of 'the artist' as currently conceived. First, the concept of performance envisages a practice which positively welcomes a process of co-creation between the various members of the development 'team'; if we can embrace this concept of performance in our work, maybe we can begin to leave off the inverted commas at last. Second, the idea of performance allows artists to take a positive view of audience involvement – for the daring, community-involvement in the process of designing the physical artifacts of urban form themselves; for the more conventional, at least designing the hardware itself around types which offer supports for the exercise of the user's own exploratory choice. Finally, if it is presented aright, the idea of performance can certainly be used to garner the cultural capital which the artist-designer needs for market survival: good actors, divas or rock musicians can earn at least as much, and are held in at least as much esteem as good architects, after all.

Overall, then, there are many advantages in re-visioning urban design as a type of co-creative performance art. Good intentions, however, are not enough; at a practical level, new tools and techniques are required if we are to perform creatively in this socio-spatial medium. First, we need new tools to forge effective co-creative links across the user–producer divide. This means that artists will have to work in ways which positively invite lay people to take advantage of whatever potential for involvement is grud-gingly made available by those who really want to exclude them. For artists, however, these attempts to express design proposals in ways and concepts which lay people can easily understand, demystifying design ideas by clarifying the reasons why particular types have been chosen and interpreted as they have, can potentially form part of the art work itself, as independent art curator Mary Jane Jacobs suggests:

> When the artistic strategies become one with the educational events, we
> have a new way of thinking about the purpose of the work. The process

that involves all of these activities needs to be recognized as the central part of the work of art. We're not just talking about a final product to which all else is preliminary. The artist him or herself as a spokesperson is a very different kind of role.[30]

Viewed in this light, the idea of the explanatory design rationale, which we discussed in the last chapter in terms of expertise, takes on a further tone, potentially far from the cold rationality of expertise itself. It calls, I think, for an attempt to create that 'familiarization of the world through laughter and popular speech', which Mikhael Bakhtin saw as indispensable for 'making possible free, scientifically knowable and artistically realistic creativity'.[31] Since no one I know of has ever done it better than Bakhtin himself, I can do no better than quote his own famous words:

> Laughter has the remarkable power of making an object come up close, of drawing it into a zone of crude contact where one can finger it familiarly on all sides, turn it upside down, inside out, peer at it from above and below, break open its external shell, look into its center, doubt it, take it apart, dismember it, lay it bare and expose it, examine it freely and experiment with it. Laughter demolishes fear and piety before an object, before a world, making of it an object of familiar contact and thus clearing the ground for an absolutely free interpretation of it.[32]

Far from destroying the essence of the creative act, as the mainstream of art culture would have it, we can see from this perspective that the skill of opening up the designer's creativity to the scrutiny of others, through the medium of a design rationale seen as art work in itself, is an indispensable part of developing an avant-garde art of public space.

As well as developing new skills for drawing users into a co-creative process, artists in the field of urban design need techniques for targeting their limited power in the most effective ways, in the face of likely power bloc ambivalence, if not outright opposition. This means that the artist here shares the expert's need for the skills to carry out financial feasibility analyses. Far from financial skills being evidence of an unhealthy, anti-art profit-orientation, as current mainstream ideology has it, they constitute knowledge which is critical to avant-garde art practice, since they are essential if designers are to push as far as they can in the face of economic opposition. As the US urban designer Jonathan Barnett put it long ago in *Urban Design as Public Policy*:

> The day-to-day decisions about the allocation of government money according to conflicting needs and different political interests, or the economics of real-estate investment, are in fact the medium of city design, as essential to the art as paint to the painter. To produce

meaningful results, from both a practical and artistic point of view, urban designers must rid themselves of the notion that their work will be contaminated by an understanding of political and real estate decisions. It is not always necessary to approve; it is essential to understand.[33]

Despite the fact that his book remains a well-respected text in urban design circles, almost a quarter of a century after it was first written, little attention has been paid by artist-designers to this financial side of Barnett's argument. This very deafness bears silent testimony to the power of the mainstream ideology of 'money is anti-art', and emphasises the point that new practical techniques will only be adopted when there exists an appropriate cultural framework, such as that which we have begun to explore in this chapter, within which they can be incorporated.

What all this suggests is that the user–producer gap, though endemic to capitalist situations, can potentially be bridged through a creative redesigning of the culture of design itself. This work of re-imagination has to encompass many levels. The art of urban design has to be prised free from its reliance on arts like painting and sculpture as models, to be reconceived as an art of performance, using as its score a repertoire of types which can support the creative desires both of the users and of the professionals who are involved in the co-creation process, at all scales from overall settlement forms to building details. Users must be re-imagined, and must re-imagine themselves, not as passive recipients but as active participants in the process of design. The art work itself has to be reconceived not as an object-building, but as an intervention affecting the serial experience of the public realm; itself re-imagined not as abstract space, but as a stage whose *raison d'être* lies in its animation through its users' own performances. All in all, everything we have discussed in this chapter leads us towards a far-reaching programme for re-imagining the culture of design-as-art, through which artists might be able to reclaim a truly avant-garde, leadership role from the current wreckage of a once-powerful tradition of art, sadly hijacked to support the status quo.

Conclusion: exciting prospects

We have now reached the end of our four-stage exploration of the processes through which urban settings are made and used. What have we gathered *en route*, to help us move forward, to get better-loved places on the ground?

During the first stage, we developed a critical ability to weigh up the pros and cons of the various ways in which the form-production process is commonly explained. On the one hand, we saw that we are not in the grip of external forces such as a disembodied 'technology' or a 'spirit of the age', but on the other hand we also came to understand that we do not have total freedom to do anything we wish. We came to realise that we make our own history, but we do so in ways which are both enabled and constrained by social structures, which are both the medium and the outcome of all our actions. In the light of this awareness, by the end of this first stage of our journey we had grasped the essential workings of the form-production process. We understood the central role of typological structures, and we saw how resources are deployed to bring about typological change. We saw that individual efforts are at the heart of this process of change, but we saw also that we become more effective agents of change to the extent that we develop alliances with others, to pool our resources; and to the extent that we develop our own internal resources of knowledge and morale. Finally, at the heart of these internal resources, we developed the critical awareness that concepts of 'good design' are always socially constructed: if we are to avoid having the wool pulled over our eyes, we have always to ask 'good for whom?'

To develop the internal knowledge resources we need, we embarked in Part II on a more detailed and concrete exploration of today's form-production process. We came to see which of the typological transformations of the capitalist era have come into good currency because they support the continued profitable working of the capital accumulation process, and we saw that working practices and concepts of 'good design' within mainstream design culture have largely developed in ways which enable well-intentioned designers to support these profitable transformations, believing them to offer users enhanced levels of choice, and support for widespread aspirations. By the end of Part II, then, we had a sufficient grasp of today's form-production process to begin to think about ways of targeting our limited power towards getting better-loved places on the ground. In Part III, we began this process by identifying a set of key experiential qualities – variety, accessibility, legibility, robustness, identity, cleanliness, biotic support and aesthetic richness – which would be widely valued by users, designers and members of the environmental lobby alike. We saw that recent urban transformations have reduced the levels of these qualities so far as many users are concerned, unless they are able to deploy large amounts of countervailing resources, with negative ecological consequences. We concluded, therefore, that there is a large potential constituency for change – to exploit, in users' interests, the new opportunity-space opened up by international political forces in the face of widespread fears of ecological crisis.

In the book's final part, we explored ways of taking up these new opportunities in practice, within the cultural potentials offered by Modernism – the imagined community to which most of today's designers feel they belong, and whose support they need if they are to find the courage to buck the current system. Within this Modernist cultural frame, we identified a typological repertoire which appears to offer the best basis for new, more user-positive and more ecologically benign transformations, together with new conceptions of 'the expert' and 'the artist' through which this repertoire could be called on in design practice.

The new conceptions of art and expertise, as we saw, have much in common: new experts and new artists would have a common concern for co-creative working with users and with other experts and artists within the form-production team itself, and for developing techniques of financial analysis which would help them to push project-orientated developers to the limits of financial feasibility. These similarities, in effect, construct a bridge across the current expert/artist divide; enabling us to envisage a new form-production culture in which the questioning, challenging thrust of art is united with the monitoring, rationalising power of expertise in a new, more integrated form of design practice. Freed from the constraints of the divisive ideologies which currently split them apart, designers of all kinds might present a united front with users in support of a sexy ecology.

In the end, then, the insights we have gained throughout this book are hopeful, inspirational ones. Looking backwards, of course, we see a gloomy picture; but looking ahead, we can see exciting prospects for change. These new prospects are not just utopian dreams. Looking ahead through the windows of opportunity we have opened, we can see that our proposals for cultural change have potential support from four key sources of power. First and most important, these proposals can attract widespread support from users themselves, since they address key problems of recent transformations which matter in users' terms. Second, this broad potential for gaining users' support opens our programme up to an alliance with the environmental lobby's global political power. Third, this global political support is the best hope we have for getting the developer on board, creating a level playing field on which profit-orientated developers can depart from current user-negative types without losing out to their competitors in the global marketplace. Finally, despite its rejection of much which is current in mainstream design culture, our new programme still has potential for gaining support from today's design professionals, because it reclaims – in the user's interests – the culture of Modernism in which so many designers find their spiritual home.

With its potential across these four complementary levels of support, this is certainly a programme which can be moved into the mainstream of the form-production process. This will only happen in practice, however, if those of us who want to see better-loved places on the ground are willing to engage in a cultural battle to influence those with access to the political and economic resources which are needed to bring change about. This is a battle in which neutrality is impossible: if we think we are neutral, then we probably support the status quo. If users make their voices heard in focused support, if politicians parlay this support to create a level playing field on which developers can depart from current types without going bankrupt, and if design professionals supply the new kinds of expertise and art which are called for, then this is a battle we can win.

Notes

Introduction

1 Ravetz, 1986, 8.

Part I: Problematics of production

1 Abrams, 1982, xv.

1 Untouched by human hand (well, almost)

1 Spengler, 1918, in Heller, 1952, 45.
2 Giedion, 1967 (1941).
3 Ambrose, 1994, 7.
4 Cited in Collins, 1965, 156.
5 Conder, 1949, 12.
6 Spengler, 1918, in Heller, 1952, 145.
7 Pevsner, 1945, 67.
8 Cited in Curtis, 1982, 264.
9 Eisenman and Gehry, 1991.
10 Heller, 1952.
11 Rogers, 1990, 21.
12 Giddens, 1981, 18.
13 Rutter, n.d., 175.
14 Abraham, in Rapoport 1969, 24.
15 Rapoport 1969, 2.
16 Banham, 1969, 288.
17 Wright, 1991, 176.
18 Ibid., 1.
19 Le Corbusier, 1946, in Lawson, 1980, 140–1.
20 Culvahouse, 1988, 37.
21 Ibid.
22 Giedion, 1934, cited in Giedion, 1967 (1941), 476.
23 Hunt, unpublished lecture at Oxford School of Architecture, 1979.
24 Gallion and Eisner, 1975, 6a.
25 Giedion, 1967 (1941).
26 For discussion of the 'house-machine', see Le Corbusier, 1946 (1923), 210–47.
27 El-Sherif, 1995.
28 Abedin, 1991.
29 Convocation Coffee House, 1990.
30 *The Times*, 25 January 1995.

31 Tschumi, 1996, 21.
32 Reported in Brown, 1992, 127.
33 Bentley *et al.*, 1985, Ch.4.
34 Hulme Regeneration Ltd, 1994.
35 Foucault, 1977; Evans, 1982.
36 Cowan, 1964.
37 Musgrove and Doidge, 1971.
38 For discussion, see Duffy, 1969, 4–13.
39 Banham, 1957.

2 Heroes and servants, markets and battlefields

1 Rand, 1943.
2 Gunts, 1993.
3 Cuff, 1991, 1.
4 Rand, 1994 (1943).
5 For discussion see Cadman and Austin Crowe, 1978.
6 Rand, op. cit., 714.
7 Pawley, 1991, 5.
8 Phillippo, 1976, 87, cited in Rabinowitz, 1996, 36.
9 Cited in Prak, 1984, 113.
10 Morton, 1992, 10.
11 Mackinnon, 1968.
12 Morton, 1990, 73.
13 Hershberger, 1969; Groat, 1982; Devlin, 1990.
14 Hubbard, 1994.
15 Morton, 1992, 11.
16 Rueschemeyer, 1986, 108.
17 Lampugnani, 1992, 114.
18 In Lipman, 1976, 24.
19 Tschumi, 1996, 20–1.
20 Ibid., 21.
21 Hadid in Noever (ed.), 1991, 51.
22 Roweis, 1988.
23 Bourdieu, 1984.
24 Hadid, op. cit., 51.
25 Eberhard, 1970, 364–5.
26 Cadman and Austin-Crowe, 1978, 208.
27 Bolte and Richter, 1965.
28 Salaman, 1970.
29 Blau, 1987.

3 Genius and tradition

1 Watkin, 1977, 115.
2 Botta, in Kloos (ed.), 1991, 28, 29 (my emphases).
3 See, for example, Polanyi, 1958; Kuhn, 1962; Gilbert and Mulkay, 1984; Latour, 1992.
4 Wynne, in Lash, Szerszynski and Wynne (eds), 1996.
5 Rosenberg, 1960.
6 Giddens, 1984.
7 Ibid.
8 Ibid., 21.
9 Ibid.

10 For discussion, see Giddens, 1984, Ch.1.
11 Therborn, 1980, 18.
12 Ibid.
13 Ibid.
14 Giddens, 1984, 17.
15 Klein, 1975.
16 Nietzsche, 1969, 162.
17 Harré, 1979, 142–3. For a useful critical discussion of Harré's ideas, see Dickens, 1990, esp. Ch. 1.
18 Castoriadis, 1992, 6.
19 Therborn, 1980, 16.
20 Elliott, 1996, 34.
21 Franck and Schneekloth, 1994, 15.
22 Therborn, 1980, 17.
23 Ibid.
24 Tschumi, 1996, 93.
25 Franck and Schneekloth, 1994, 9.
26 Cited in Vidler, 1989, 152.
27 Hillier and Leaman, 1974, 4.
28 Franck and Schneekloth, 1994, 16.
29 Heyer, 1966, 392.
30 Ibid.
31 Stravinsky, 1970, 75.
32 Lawson, 1980.
33 de Bono, 1970, 9.
34 Ibid.
35 Wright, 1949.
36 March and Steadman, 1974.
37 Bandini, 1993, 394.
38 Lawson, 1980, 110.
39 Deleuze and Guattari, 1994, 25.
40 Ibid.
41 Grosz, 1994, 165.
42 Therborn, 1980, 78.
43 Elliott, 1996, 34.
44 Collins, 1965, 277.
45 Venturi, 1996, 274.

Part II: Spatial transformations and their cultural supports

1 Young, C. in Gottdiener and Pickvance (eds), 1991, 202.
2 For discussion, see Offe, 1985, and Lash and Urry, 1987.
3 For discussion at a cultural level see, for example, Hall in Storey, 1994, 465.

4 Profit and place

1 *OED*, 1989, 863.
2 Ambrose, 1986, 7.
3 Lefebvre, 1991.
4 See, for example, Mandel, 1969, 25–6.
5 For discussion, see Rittel, 1976, 77–91.
6 Turner, 1926, 126.
7 Wolfe, 1992, 78–9.
8 For an interesting discussion of the simplification of detailed drawings during

the late nineteenth century in the USA, see Cardinal-Pett in Rüedi *et al.* (eds), 1996.
9 Cornes, 1905, xiv.
10 Goodhart-Rendel, 1935, cited in Davey, 1995, 142.
11 For discussions of cities as movement economies, see Hillier and Hanson, 1984, and Hillier, 1996.
12 Sitte, 1965 (1901 edn), 93 (my emphases).
13 van Brunt, 1969 (1886).
14 For discussion related to this point, see Harvey, 1985, 22.

5 Propping up the system

1 Mandel, 1970, 51.
2 Campbell, 1987, 85.
3 Aron, 1968, 64.
4 For discussion, see Aglietta, 1987, 129–30.
5 Engels, 1971, 55.
6 Stübben, 1890, 217, cited in Angelil, 1991, 18.
7 Hillier and Hanson, 1984.
8 Sennett, 1973, 55–6.
9 Zola, 1883.
10 Reeve, 1995.

6 Building bastions of sense

1 For example, in the UK context, see MORI, 1989.
2 Therborn, 1980, 19.
3 Hall, 1995.
4 Berger, 1971, 18.
5 Ibid., 32.
6 Wolfe, 1993.
7 In Anthony, 1991, 41.
8 Weston, 1996, 7.
9 Wolfe, 1993.
10 *Interbuild* 6 (1959), 9–11, cited in Prak, 1984, 94.
11 *RIBA Journal*, 1965, cited in Lasdun, 1980, 120.
12 Pearman, 1991, 5.13.
13 Rand, 1944 (1943).
14 Cited in Bouman and Van Toorn, 1994, 55.
15 Pichler, W. and Hollein, H., 'Absolute Architektur', 1962, 175, quoted in Prak, 1984, 134.
16 Goodwin, quoted in Marder (ed.), 1985, 23.
17 For discussion, see Stamps, 1994; Myers and Myers, 1980; and McKinnon, 1962.
18 Battersby, 1989.
19 Cited in White, 1970, 104 (mix of upper and lower case in original advertising copy removed for ease of reading).
20 Butler, 1962, cited in Parker and Pollock, 1981, 7.
21 Gabhart and Broun, 1972, cited in Parker and Pollock, 1981, 6.
22 Howard, 1898.
23 For discussion, see Bentley, 1981.
24 Garreau, 1991.
25 Stull, 1975, 353, cited in Lai, 1994, 86.
26 Stull, 1975.

27 Newman, 1972.
28 Coleman, 1985.
29 This campaign produced a range of pamphlets and guidance notes too numerous and ephemeral to list.
30 Tarn, 1973.
31 Le Corbusier, 1929.
32 Greater London Council, 1978.
33 In Teymur *et al.* (eds), 1988, 63.
34 Sitte, 1965.
35 Cullen, 1971.
36 Giedeon, 1967 (1941), 436.
37 Ibid., 879.
38 Pevsner, 1979.
39 Le Corbusier, 1946 (1923), 44.
40 McMillan, 1979.
41 Cited in Banham, 1957, 86.
42 Tschumi, 1996, 17.
43 Glancey, 1989, 27.
44 Venturi, 1996, 310.
45 Ibid., 262.
46 Weston, 1996, 230.
47 Sola Morales, 1989.
48 Morton, 1992, 3.

Part III: Positive values, negative outcomes

1 Jowell, 1994, 135.

7 Concepts for prospecting common ground

1 Jowell, 1994, 135.
2 Ross, 1994, 15.
3 Ibid., 17.
4 Miller, 1991, 8.
5 Ibid.
6 I am indebted for this example to Binoy Panicker.
7 See, for example, Collins, 1990.
8 Kennedy and Piette, in Aaron, J. and Walby, S. (eds), 1991, 33.
9 Miller, 1991, 8.
10 de Certeau, 1984, 31.
11 Lefebvre, 1991.
12 Gibson, 1979.
13 Thrift, 1996, 43.
14 See, for example, Hershberger, 1969; Groat, 1982; Devlin, 1990; Hubbard, 1994.
15 For discussion, see Baudrillard, 1981.
16 Bauman, 1991, 170.
17 Ibid.
18 Inglis, 1993, 179.
19 Dahrendorf, 1979.
20 Featherstone, 1991, Ch. 5.
21 Farmer, 1996.
22 Anderson, 1991.
23 Bauman, 1991, xix (emphases in the original).

24 Whyte, 1980.
25 Oosterman, 1993, 299.
26 Ibid.
27 Home Office Research and Planning Unit, 1994.

8 Beyond buzzwords

1 Bauman, 1991.
2 See, for example, Goffman, 1967.
3 Interview with Dora Boatemah, Project Director of Angell Town Community Project Ltd, 15 January 1994.
4 Raban, 1975, 184.
5 Bureau of the Census of the United States, 1993.
6 Cited in Flores, 1996, 20.
7 Appleyard and Lintell, 1975.
8 Ibid., 10
9 Ibid.
10 Oostermann, 1993, 299.
11 Ibid.
12 Neponsechic, 1997.
13 Ibid. 30.
14 Vasak, 1989.
15 Joint Centre for Urban Design, 1996.
16 For interesting discussion of the 'gated community' approach, see Davis, 1990.
17 Poyner and Webb, 1991.
18 Washbrook, 1994.
19 Hillier, 1988, 63–88.
20 Ibid, 86.
21 Oxford Brookes University Urban Regeneration Consultancy, 1989.
22 Cited in Klodawsky and Lundy, 1994, 130.
23 Bosley, 1993.
24 Bennett and Wright, 1984.
25 Home Office Research and Planning Unit, 1992.
26 For example, see Tsoskonoglou, 1995.
27 Lea and Young, 1993.
28 Elsom, 1996, 1.
29 Ibid., 2.
30 Hillier and O'Sullivan, 1989.
31 Allied Dunbar, Health Education Authority and Sports Council, 1992.
32 Morris *et al.*
33 Armstrong, 1993, 41.
34 Southwood, 1961, cited in Hough, 1989, 197.
35 I am indebted for this example, from his ongoing PhD research, to Danny O'Hare of Queensland University of Technology.
36 Rapoport, 1977.
37 Oskam, 1995, 5.

9 Horizons of choice

1 Interview with Dora Boatemah, Project Director of Angell Town Community Project Ltd, 11 June 1994.
2 From pre-course notes submitted by a planner from a British Midland town, 1995.
3 For a full discussion, see Hillier and Hanson, 1984.

4 Hillier, in Teymur *et al.* (eds), 1988, 77.
5 Help the Aged, 1996.
6 Department of Transport, 1995.
7 Dunning, 1997.
8 Townsend, 1984, 29.
9 For discussion, see Lynch, 1960, Ch. III.
10 Ibid., 47.
11 Ibid., 31.
12 Ibid., 106.
13 Alcock, in Hayward and McGlynn (eds), 1993.
14 Hillier, 1992, 44.
15 Ibid.
16 Lynch, 1960, 105.
17 Al-Moayyed *et al.*, 1994, 56.
18 *The Times*, 15 October 1995, 3.3.
19 Ibid.
20 Zwarts, M., article in Kloos, M. (ed.), 1991, 151.

Part IV: Windows of opportunity

10 Reclaiming the Modernist vision

1 For an earlier exploration of how to use some of these types, see Bentley *et al.*, 1985.
2 Abedin, 1991.
3 Ko, 1998.
4 For discussion, see Polano, 1988.
5 Buch, 1994, 104.
6 Cited in Curtis, 1996, 246.
7 In Kloos, 1988, 18.
8 In Kloos, 1988, 52.
9 In Kloos, 1988, 64.
10 Cited in Curtis, 1996, 246.
11 Jacobs, 1961.
12 For a succinct account of these processes, see Couch, 1990.

11 Experts who deliver

1 Bloomer, 1996, 14.
2 Ibid., 18.
3 Fiske, 1989, 31.
4 MORI, 1989.
5 Oxford Brookes University Urban Regeneration Consultancy, 1989.
6 Wynne, 1996.
7 Ibid., 67.
8 Slim and Thompson, 1993, 7–8.
9 Greed and Roberts, 1998, 9.
10 *Architects' Journal*, Vol. 198, No.5, 1993, pp. 27–38.
11 http:/rudi.herts.ac.uk/cs/angell/angell_town.htm
12 Samuels, 1993, 120.

12 Artists in a common cause

1 Inglis, 1993, 183.

2 hooks, 1995.
3 Bourdieu and Haacke, 1995, 2.
4 Ibid.
5 Ibid., 184.
6 Interview with Councillor David Brown, 15 November 1994.
7 Interview with Dr Dora Boatemah, 23 September 1990.
8 Weissman, 1980, cited in Bourdieu and Haacke, 1995, 8.
9 Tschumi, 1996, 14.
10 Storey, 1994, 232.
11 Ibid., 235.
12 Eagleton, 1990, 372.
13 Discussed in Schröter, 1959, 223 ff.
14 For a useful general discussion of 'new genre public art', see Lacy, 1995.
15 SPARC, 1994, 12.
16 Grout and Palisca, 1996.
17 Asplund, 1995, 424, 430.
18 Rowe, 1995.
19 For discussion see 'Treason of the Intellectuals? Benda, Benn and Brecht' in Timms and Collier, 1988.
20 For a stimulating account of urban space as public stage, see Jarvis, 1996.
21 Jacobs, 1961.
22 For a brief introduction to the work of Anna Halprin, see Lacy, 1995, 228–9. For Lawrence Halprin, see Halprin, 1966 and 1969.
23 Eisenstein, 1991, 59.
24 Eisenstein, 1960, 170.
25 Bourdieu, 1984.
26 Rattenbury, 1994, 35.
27 Tschumi, MIT, 1996, 17.
28 Cited in Lacy, 1995, 44–5.
29 Ibid.
30 Cited in Lacy,1995, 40.
31 Bakhtin, 1990 (1981), 23.
32 Ibid.
33 Barnett, 1974, 6.

Bibliography

Aaron, J. and Walby, S. (eds), *Out of the Margins*, London, Falmer, 1991.

Abedin, Z., *Islamic Methods in Urban Design*, unpublished MA dissertation, Oxford Brookes University, Joint Centre for Urban Design, 1991.

Abraham, R. J., *Elementare Architektur*, Salzburg, Residentz Verlag, n.d.

Abrams, P.A., *Historical Sociology*, Shepton Mallet, Open Books, 1982.

Adam, R., *Classical Architecture*, London, Viking, 1990.

Aglietta, M., *A Theory of Capitalist Regulation, the US Experience*, London, NLB, 1987.

Alcock, A., 'Aesthetics and Urban Design' in Hayward and McGlynn, 1993.

Allied Dunbar, Health Education Authority and Sports Council, *National Fitness Survey*, Allied Dunbar, 1992.

Al-Moayyed, M. *et al.*, *Users' Demands in the Urban Design Process*, unpublished Issues project, Oxford, Joint Centre for Urban Design, 1994.

Ambrose, P., *Whatever Happened to Planning?*, London, Methuen, 1986.

Ambrose, P., *Urban Process and Power*, London, Routledge, 1994.

Anastasi, A., *Psychological Testing*, 2nd edn, New York: Macmillan, 1961.

Anderson, B., *Imagined Communities, Reflections on the Origin and Spread of Nationalism*, London, Verso, 1991.

Angelil, M. (ed.), *On Architecture, the City and Technology*, Boston, Mass., Association of Collegiate Schools of Architecture, 1991.

Anthony, K. H., *Design Juries on Trial: The Renaissance of the Design Studio*, New York, Van Nostrand Reinhold, 1991.

Appleyard, D. and Lintell, M., 'Streets: Dead or Alive', *New Society*, 3 July 1975.

Armstrong, N., 'Independent Mobility and Children's Physical Development', in Hillman, M. (ed.), 1993.

Arnell, P. and Bickford, T. (eds), *James Stirling*, New York, Rizzoli, 1984.

Aron, R., *Progress and Disillusion: The Dialectics of Modern Society*, London, Pall Mall, 1968.

Asplund, C., 'Frederic Rzewski and Spontaneous Political Music', in *Perspectives of New Music*, Vol. 33, No.2, Summer 1995.

Baca, J. F., 'Whose Monument Where? Public Art in a Many-cultured Society', in Lacy, S. (ed.) 1995.

Bakhtin, M., 'Epic and Novel', in Holquist (ed.) 1990 (1981).

Bandini, M., 'Typological Theories in Urban Design', in Farmer and Louw (eds) 1993.

Banham, R., 'Ornament and Crime, the Decisive Contribution by Adolf Loos', *Architectural Review*, Vol.121, February 1957.

Banham, R., *The Architecture of the Well-Tempered Environment*, London, Architectural Press, 1969.

Barnett, J., *Urban Design as Public Policy*, New York, McGraw Hill, 1974.

Battersby, C., *Gender and Genius: Towards a Feminist Aesthetics*, London, Women's Press, 1989.

Baudrillard, J., *For a Critique of the Political Economy of the Sign*, St Louis, Telos, 1981.

Bauman, Z., *Intimations of Postmodernity*, London, Routledge, 1991.

Bennett, T. and Wright, R., *Burglars on Burglary*, Aldershot, Gower, 1984.

Bentley, I., 'Individualism or Community', in Oliver *et al.*, 1981.

Bentley, I. *et al.*, *Responsive Environments*, London, Architectural Press, 1985.

Berger, P., *A Rumour of Angels: Modern Society and the Rediscovery of the Supernatural*, Harmondsworth, Penguin, 1971.

Blau, J. R., *Architects and Firms: A Sociological Perspective on Architectural Practice*, Cambridge, Mass., MIT Press, 1987.

Bloomer, J., 'The Matter of the Cutting Edge' in Rüedi, K. *et al.* (eds), *Designing Practices*, London, Black Dog, 1996, 18.

Bohigas, O., untitled in Kloos (ed.), 1988.

Bolte, K. M. and Richter, H. J., 'Der Architekt: sein Beruf und seine Arbeit', *Detail*, Vol.8, No.5, 1965.

Bono, E. de, *The Dog-Exercising Machine: A Study of Children as Inventors*, Harmondsworth, Penguin, 1970.

Bosley, T., *The Perception of Safety in Public Space*, unpublished MA dissertation, Oxford Brookes University, Joint Centre for Urban Design, 1993.

Bouman, O. and Van Toorn, R., *The Invisible in Architecture*, London, Academy Editions, 1994.

Bourdieu, P., *Distinction, A Social Critique of the Judgement of Taste*, trans. Nice, R., London, Routledge and Kegan Paul, 1984.

Bourdieu, P., *Sociology in Question*, London, Sage, 1993.

Bourdieu, P. and Haacke, H., *Free Exchange*, Cambridge, Polity Press, 1995.

Brown, F., 'Building Transformations: Commentary', in *Proceedings of IAPS 12 Conference, Thessaloniki, Aristotle University, 1992*, Vol. 3.

Brunt, A. van, 'On the Present Condition and Prospects of Architecture' (1886) cited in Coles, 1969.

Buch, J., *A Century of Architecture in the Netherlands 1880/1990*, Amsterdam: Netherlands Architecture Institute, 1994.

Bureau of the Census of the United States, *American Housing Survey*, 1993.

Burke, J. and Caldwell, C., *Hogarth: The Complete Engravings*, London, Thames and Hudson, 1968.

Cadman, D. and Austin Crowe, L., *Property Development*, London, Spon, 1978.

Campbell, C., *The Romantic Ethic and the Spirit of Modern Consumerism*, Oxford, Blackwell, 1987.

Cardinal-Pett, C., 'Detailing', in Rüedi *et al.* (eds), 1996.

Castoriadis, C., 'Reflections on Racism', in *Thesis Eleven*, 1992.

Certeau, M. de, *The Practice of Everyday Life*, Berkeley, University of California, 1984.

Cloke, P. (ed.), *Policy and Change in Thatcher's Britain*, Oxford, Pergamon, 1992.

Cohen, M. *et al.*, *Preparing for the Urban Future: Global Pressures and Local Forces*, Baltimore, Johns Hopkins University Press, 1996.

Coleman, A., *Utopia on Trial: Vision and Reality in Planned Housing*, London, Shipman, 1985.

Colenbrande, B. (ed.), *Style Standard and Signature in Dutch Architecture*, Rotterdam, NAI, 1993.

Coles, W. A., *Architecture and Society: Selected Essays of Henry van Brunt*, Cambridge, Mass., MIT Press, 1969.

Collins, P., *Changing Ideals in Modern Architecture, 1750–1950*, London, Faber and Faber, 1965.

Collins, P. H., *Black Feminist Thought*, London, Harper Collins, 1990.

Conder, N., *An Introduction to Modern Architecture*, London, Art and Technics, 1949.

Convocation Coffee House, Oxford, *Menu*, 1990.

Cornes, J., *Modern Housing in Town and Country*, London, Batsford, 1905.

Couch, C., *Urban Renewal, Theory and Practice*, London, Macmillan, 1990.

Cowan, P., 'Studies in the Growth, Change and Ageing of Buildings', *Transactions of the Bartlett Society*, 1, London, Bartlett School of Architecture, 1964.

Crowther, P., *Critical Aesthetics and Post Modernism*, Oxford, Clarendon Press, 1993.

Cuff, D., *Architecture: The Story of Practice*, Cambridge, Mass., MIT Press, 1991.

Cullen. G., *Townscape*, London, Architectural Press, 1961.

Cullen, G., *The Concise Townscape*, London, Architectural Press, 1971.

Culvahouse, T., 'Figuration and Continuity in the Work of H. H. Richardson', in *Perspecta 24*, New York, Rizzoli, 1988.

Curtis, W. J. R., *Modern Architecture since 1900*, London, Phaidon, 1982.

Dahrendorf, R., *Life Chances, Approaches to Social and Political Theory*, London, Weidenfeld and Nicholson, 1979.

Davey, P., *Arts and Crafts Architecture*, London, Phaidon, 1995.

Davis, M., *City of Quartz*, London, Vintage, 1990.

Deleuze, G. and Guattari, F., *Anti-Oedipus, Capitalism and Schizophrenia*, London, Athlone, 1984.

Deleuze, G. and Guattari, F., *What is Philosophy?*, London, Verso, 1994.

Department of Transport, *Transport Statistics Great Britain*, London, HMSO, 1995.

Devlin, K., 'An Examination of Architectural Interpretation: Architects Versus Non-Architects' *Journal of Architectural and Planning Research*, 7 (3), 235–44, 1990.

Dickens, P., *Urban Sociology*, Hemel Hempstead, Harvester Wheatsheaf, 1990.

Duffy, F., 'Role and Status in the Office', *A. A. Quarterly*, Vol. 1, No. 4, October 1969.

Duivesteijn, A., untitled, in Colenbrande, B. (ed.), 1993.

Dunning, J., 'Rising Car Ownership among Women and the Elderly Continues to Fuel Growth in Transport Demand', *Local Transport Today*, 202, 3 January 1997, 9.

Eagleton, T., *The Ideology of the Aesthetic*, Oxford, Blackwell, 1990.

Eberhard, J. P., 'We Ought to Know the Difference', in Moore (ed.), 1970.

Eisenstein, S., *Nonindifferent Nature*, trans. Marshall, H., Cambridge, CUP, 1960.

Eisenstein, S., 'Montage and Architecture', trans. Glenny, M., in Glenny and Taylor (eds), 1991.

Elliott, A., *Subject to Ourselves: Social Theory, Psychoanalysis and Postmodernity*, Cambridge, Polity Press, 1996.

El-Sherif, A., *Reconciling Traditions and Contemporary Aspirations in the Arab-Islamic*

City: Alexandria, Egypt, unpublished PhD, Oxford, JCUD, Oxford Brookes University, 1995.

Elsom, D., *Smog Alert: Managing Urban Air Quality,* London, Earthscan, 1996.

Engels, F., *The Condition of the Working Class in England,* trans. and ed. Henderson, W. O. and Chaloner, W. H., Oxford, Blackwell, 1971.

Evans, R., *The Fabrication of Virtue,* Cambridge, CUP, 1982.

Farmer, B. and Louw, H. (eds), *Companion to Contemporary Architectural Thought,* London, Routledge, 1993.

Farmer, J., *Green Shift: Towards a Green Sensibility in Architecture,* Oxford, Butterworth-Heinemann, 1996.

Featherstone, M., *Consumer Culture and Postmodernism,* London, Sage, 1991.

Fiske, J., *Understanding Popular Culture,* London, Routledge, 1989.

Flores, E. O., 'Towards a city of solidarity and citizenship', *Environment and Urbanization,* Vol.8, No.1, April 1996.

Fogarty, F., 'How Today's Clients Pick Architects', *Architectural Forum,* 110, pp. 114–15, 1959.

Foucault, M., *Discipline and Punish: the Birth of the Prison,* London, Allen Lane, 1977.

Franck, K. A. and Schneekloth, L. H. (eds), *Ordering Space: Types in Architecture and Design,* New York, Van Nostrand Reinhold, 1994.

Gabhart, A. and Broun, E., *Walters Art Gallery Bulletin,* Vol.24, No.7, 1972.

Gallion, A. B. and Eisner, S., *The Urban Pattern: City Planning and Design,* New York, Van Nostrand, 1975.

Garreau, J., *Edge City: Life on the New Frontier,* New York, Doubleday, 1991.

Gartman, D., *Auto Opium: A Social History of American Automobile Design,* London, Routledge, 1994.

Gasparini, G. and Posani, J. P., *Caracas, A Traves de su Arquitectura,* Caracas, Fina Gómez, 1969.

Gibson, J. J., *The Ecological Approach to Visual Perception,* Boston, Houghton Mifflin, 1979.

Giddens, A. *A Contemporary Critique of Historical Materialism,* Vol.2, *Power, Property and the State,* London, Macmillan, 1981.

Giddens, A., *The Constitution of Society: Outline of a Theory of Structuration,* Cambridge, Polity Press, 1984.

Giddens, A., 'Elements of the Theory of Structuration', in Giddens, A., *et al.* (eds), 1994.

Giddens, A., *et al.* (eds), *The Polity Reader in Social Theory,* Cambridge, Polity Press, 1994.

Giedion, S., 'Nouveaux Ponts de Maillart', *Cahiers d'Art,* Vol.IX, 1934, Nos 1–4.

Giedeon, S., *Space, Time and Architecture,* Cambridge, Mass., Harvard University Press, 1967 (1941).

Gilbert, N. and Mulkay, M., *Opening Pandora's Box: A Sociological Analysis of Scientists' Discourse,* Cambridge, Cambridge University Press, 1984.

Glancey, J., *New British Architecture,* London, Thames and Hudson, 1989.

Glenny, M. and Taylor, R. (eds), *Goethe, Selected Works,* Vol.2, London, British Film Institute, 1991.

Goffman, E., *Interaction Ritual,* New York, Pantheon, 1967.

Goodhart-Rendel, H. S., 'The Work of Beresford Pite and Halsey Ricardo', *RIBA Journal,* Vol.XLIII, 1935.

Goodwin, B., 'Architecture of the Id', *Architecture and Urbanism*, No.117, 1980.

Gottdiener, M. and Pickvance, C. G. (eds), *Urban Life in Transition*, London, Sage, 1991.

Greater London Council, *Introduction to Housing Layout*, London, Architectural Press, 1978.

Greed, C. and Roberts, M. (eds), *Introducing Urban Design*, Harlow, Longman, 1998.

Groat, L., 'Meaning in Post-Modern Architecture: An Examination Using the Methodological Sorting Task', *Journal of Environmental Psychology*, 2 (1), 3–22, 1982.

Grosz, E., *Volatile Bodies: Towards a Corporeal Feminism*, Bloomington, Indiana University Press, 1994.

Grout, D. J. and Palisca, C. V., *A History of Western Music*, New York: Norton, 1996.

Gunts, E. 'The Fountainhead at 50', *Architecture*, 82/5, 1993, 35–7.

Hadid, Z., 'Recent Work', in Noever (ed.), 1991.

Hall, P., 'The European City – Past and Future', *Proceedings: The European City – Sustaining Urban Quality*, Copenhagen, Danish Academy of Architecture, 1995.

Hall, S., 'Notes on Reconstructing "the Popular"', in Storey, 1994.

Halprin, L., *Freeways*, New York, Reinhold, 1966.

Halprin, L., *The RSVP Cycles, Creative Processes in the Human Environment*, New York, Braziller, 1966, 1969.

Harré, R., *Social Being*, Oxford, Blackwell, 1979.

Harvey, D., *Consciousness and the Urban Experience*, Oxford, Blackwell, 1985.

Hayward, R. and McGlynn, S. (eds), *Making Better Places, Urban Design Now*, Oxford, Butterworth Architecture, 1993.

Heller, H., *The Disinherited Mind: Essays in Modern German Literature and Thought*, Cambridge, Bowes and Bowes, 1952.

Help the Aged, *Licenced to Drive at 85?*, London, Help the Aged, 1996.

Hershberger, R. C., *A Study of Meaning in Architecture*, in Sanoff and Cohn (eds), 1969.

Heyer, P., *Architects on Architecture*, New York, Walker, 1966.

Hillier, B., 'Against Enclosure', in Teymur *et al.* (eds), 1988.

Hillier, B., 'Look Back to London', *Architects' Journal*, 15 April 1992.

Hillier, B. and Leaman, A., 'How is Design Possible?', *Journal of Architectural Research and Teaching*, Vol.1, No.1, 1974.

Hillier, B. and Hanson, J., *The Social Logic of Space*, Cambridge, CUP, 1984.

Hillier, B. and O'Sullivan, P., *Urban Design and Climate Change*, Proceedings of Conference C59, London, Solar Energy Society, 1989.

Hillman, M. (ed.), *Children's Transport and the Quality of Life*, London, Policy Studies Institute, 1993.

Hitchcock, Henry-Russell and Johnson, P., *The International Style*, New York: Norton, 1966 (1932).

H. M. Inspectorate of Police, *What Price Policing? A Study of Efficiency and Value for Money in the Police Service*, London, HMSO, 1998.

Holquist, M. (ed.), *The Dialogic Imagination*, Austin, University of Texas Press, 1990 (1981).

Home Office Research and Planning Unit, *British Crime Survey*, London, HMSO, 1992.

Home Office Research and Planning Unit, *British Crime Survey*, London, HMSO, 1994.

hooks, b., *Art on My Mind: Visual Politics*, New York, The New Press, 1995.

Hough, M., *City Form and Natural Process*, London, Routledge, 1989.

Howard, E., *Garden Cities of Tomorrow*, Eastbourne, Attic, 1985 (1898).

Hubbard, P. J., *Diverging Evaluations of the Built Environment: Planners Versus the Public*, in Neary, Symes and Brown (eds), 1994.

Huet, B., untitled in Kloos (ed.), 1988.

Hulme Regeneration Ltd, *Rebuilding the City: A Guide to Development, Hulme, Manchester*, Hulme Regeneration Ltd, 1994.

Hunt, A., unpublished lecture at Oxford School of Architecture, 1979.

Inglis, F., *Cultural Studies*, Oxford: Blackwell, 1993.

Jacobs, J., *The Death and Life of Great American Cities*, New York, Random House, 1961.

Jarvis, B., 'Mind/Body = Space/Time = Things/Events', *Urban Design Studies*, Vol.2, 1996, 101–14.

Joint Centre for Urban Design, *Quality in Town and Country: An Analysis of Responses to the Discussion Document*, London, Department of Environment, 1994.

Jowell, R. *et al.*, *British Social Attitudes, the 11th report*, Hampshire, Dartmouth, 1994.

Kennedy, M. and Piette, B., 'From the Margins to the Mainstream: Issues Around Women's Studies on Adult Education and Access Classes', in Aaron and Walby (eds), 1991.

Kirschenmann, J. and Muschalek, C., *Residential Districts*, New York, Watson-Guptill, 1980.

Klein, M., *The Psycho-Analysis of Children*, London, Hogarth, 1975.

Klodawsky, F. and Lundy, C., 'Women's Safety in the University Environment', *Journal of Architectural and Planning Research*, 11.2 (Summer 1994), 130.

Kloos, M. (ed.) *Amsterdam, an Architectural Lesson*, Amsterdam, Thoth, 1988.

Kloos, M. (ed.), *Architecture Now*, Amsterdam, Architectura and Natura, 1991.

Knorr-Cetina, K. and Cicourel, A.V. (eds), *Advances in Social Theory and Methodology*, Routledge and Kegan Paul, 1981.

Knox, P. L., *The Design Professions and the Built Environment*, London, Croom Helm, 1988.

Ko, A. K. P., *A Feng Shui Approach to Urban Design*, unpublished MA dissertation, Oxford Brookes University, Joint Centre for Urban Design, 1998.

Krier, L., 'Berlin-Tegel', in *Leon Krier, Houses, Palaces, Cities*, London, Architectural Design Profile, 54, 1984.

Kuhn, T. S., *The Structure of Scientific Revolutions*, Chicago, University of Chicago Press, 1962.

Lacy, S. (ed.), *Mapping the Terrain, New Genre Public Art*, Seattle, Bay Press, 1995.

Lai, W. C. L., 'The Economics of Land-Use Zoning: A Literature Review and Analysis of the Work of Coase', in *Town Planning Review*, Vol.65, pp. 77–98, 1994.

Lasdun, D., *Architecture in an Age of Scepticism: A Practitioner's Anthology*, London, Heinemann, 1984.

Lash, S. and Urry, J., *The End of Organized Capitalism*, Cambridge, Polity Press, 1987.

Lash, S., Szerszynski, B. and Wynne, B., *Risk, Environment and Modernity*, London, Sage, 1996.

Latour, B., *We Have Never Been Modern*, London, Harvester Wheatsheaf, 1992.

Lawson, B., *How Designers Think*, London, Architectural Press, 1980.

Le Corbusier, trans. Etchells, F., *Towards a New Architecture*, London, Architectural Press, 1946 (1923).

Le Corbusier, trans. Etchells, F., *The City of Tomorrow and its Planning*, London, Architectural Press, 1971 (1924).

Lea, J. and Young, J., *What Is To Be Done About Law and Order?*, London, Pluto Press, 1993.

Lefebvre, M., *The Production of Space*, Oxford, Blackwell, 1991.

Leupen, B., Deen, W. and Graffe, C. (eds), *Hoe Modern is de Nederlandse Architectuur?*, Rotterdam, NAI, 1990.

Lynch, K., *The Image of the City*, Cambridge, Mass., MIT, 1960.

MacKinnon, D., 'The Nature and Nurture of Creative Talent', *American Psychologist*, 7/171, 1962.

MacKinnon, D., 'The Personality Correlates of Creativity', in Nielson, G. S. (ed.), 1962.

Magnano Lampugnani, V., 'The Architect as Client', *Lotus International*, No.70, 1991.

Mandel, E., *An Introduction to Marxist Economic Theory*, New York, Pathfinder Press, 1969.

Mandel, E., *Marxist Theory of Alienation*, New York, Pathfinder Press, 1970.

March, L. and Steadman, P., *The Geometry of Environment*, London, Methuen, 1974.

Marder, T. I. (ed.), *The Critical Edge: Controversy in Recent American Architecture*, Cambridge, Mass., MIT Press, 1985.

McMillan, R. J., *Environmental Design Synthesis of Architecture and the Inherent Structure*, New York, Vantage Press, 1979.

Miller, D., *Material Culture and Mass Consumption*, Oxford, Blackwell, 1991.

Ministry of Transport, *Traffic in Towns*, London, HMSO, 1963.

Moore, C. T. (ed.), *Emerging Methods in Environment Design and Planning*, Cambridge, Mass.: MIT Press, 1970.

MORI, *Public Attitudes to Architecture*, unpublished research for *Sunday Times*, London, 1989.

Morris, J. N. *et al.*, 'Exercise in Leisure Time: Coronary Attack and Death Rates', in *British Heart Journal*, Vol. 63, 325–34.

Morton, R., *The Teaching of Economics in Schools of Architecture*, London, RIBA, 1990.

Morton, R., 'Professional Ideologies and the Quality of the British Environment', in *Proceedings of the Bartlett International Summer School*, London, London University, 1992.

Musgrove, J. and Doidge, C., 'Room Classification', *Architectural Research and Teaching*, Vol. 1, No. 1, May 1970.

Myers, I. and Myers, P., *Gifts Differing*, Palo Alto: Consulting Psychologists Press, 1980.

Neary, S. J., Symes, M. S. and Brown, F. E., *The Urban Experience: A People–Environment Perspective*, London, Spon, 1994.

Nielson, G. S. (ed.), *Proceedings of XIV International Congress of Applied Psychology*, Copenhagen, Munksgaard, 1962.

Noever, P. (ed.), *Architecture in Transition: Between Deconstruction and New Modernism*, Munich, Prestel, 1991.

Neponsechic, M. R., 'Unacceptable Echoes', *Places*, 11.1, 1997, 29–30.

Newman, O., *Defensible Space, Crime Prevention Through Environmental Design*, New York, Macmillan, 1972.

Nietzsche, F., *On the Genealogy of Morals*, trans. Kaufmann, W. and Hollingdale, R. J., New York, Vintage, 1969.

Offe, C., *Disorganized Capitalism*, Cambridge, Polity Press, 1985.

OED (*Oxford English Dictionary*), 2nd edn, Vol II, 1989, p. 863.

Oliver, P., Davis, I. and Bentley, I., *Dunroamin: The Suburban Semi and its Enemies*, London, Barrie and Jenkins, 1981.

Oostermann, J., *Parade der Passanten: de stad hetveiter en de terrase*, Utrecht, Jan van Arkel, 1993.

Oosterman, A., *Housing in the Netherlands*, Rotterdam, NAI, 1996.

O'Riordan, T., 'The Environment', in Cloke, 1992.

Oskam, A. W., 'A Tale of Two Cities, I: Amsterdam', *Proceedings of The European City – Sustaining Urban Quality*, Copenhagen, Danish Academy of Architecture, 1995.

Oxford Brookes University Urban Regeneration Consultancy, *Angell Town Regeneration Survey*, 1989.

Parker, R. and Pollock, G., *Old Mistresses: Women, Art and Ideology*, London, Routledge, 1981.

Paulsson, T., *Scandinavian Architecture*, London, Leonard Hill, 1958.

Pawley, M., *Terminal Architecture*, London, Reaktion, 1998.

Pearman, H., *The Times*, 25 October 1991, 5.3.

Pell, D. J., 'Local Management of Planet Earth', *Sustainable Development*, Vol.4, No.3, December 1996, 141.

Pevsner, N., *The Leaves of Southwell*, London, Penguin, 1945.

Pevsner, N. A., *A History of Building Types*, London, Thames and Hudson, 1979.

Phillippo, G., *The Professional Guide to Real Estate Development*, New York, Dow Jones, 1976.

Polano, S., *Hendrik Petrus Berlage: Complete Works*, Sevenoaks, Butterworth Architecture, 1988.

Polanyi, M., *Personal Knowledge*, London, Routledge and Kegan Paul, 1958.

Poyner, B. and Webb, R., *Crime Free Housing*, Oxford, Butterworth, 1991.

Prak, N. L., *Architects: The Noted and the Ignored*, Chichester, Wiley, 1984.

Pred, A. and Watts, M., *Reworking Modernity: Capitalisms and Symbolic Discontent*, New Brunswick, Rutgers, 1992.

Raban, J., *Soft City*, London, Fontana, 1975.

Rabinowitz, H., 'The Developer's Vernacular: The Owner's Influence on Building Design', *Journal of Architectural and Planning Research*, Vol.13, No.1, 1996.

Rand, A., *The Fountainhead*, Harmondsworth, Penguin, 1994 (1943).

Rapoport, A., *House Form and Culture*, Englewood Cliffs, NJ, Prentice-Hall, 1969.

Rapoport, A., *Human Aspects of Urban Form: Towards a Man-environment Approach to Urban Form*, Oxford, Pergamon, 1977.

Rattenbury, K., 'Echo and Narcissus', in *Architecture and Film*, AD Profile 112, London, Academy Editions, 1994.

290   *Bibliography*

Ravetz, A., *The Government of Space: Town Planning in Modern Society*, London, Faber, 1986.

Reeve, A., *Urban Design and Places of Spectacle*, Oxford Brookes University, unpublished PhD thesis, 1995.

RIBA Journal, Vol.72, No.4, April 1965.

Rittel, H., 'Evaluating Evaluators', in *Papers Published for the Accreditation Evaluation Conference of the National Architecture Accrediting Board*, New Orleans, 1976.

Rocheleau, P., Sprigg, J. and Larkin, P., *Shaker Built*, London, Thames and Hudson, 1994.

Rogers, R., *Architecture: A Modern View*, London, Thames and Hudson, 1990.

Rosenberg, H., *The Tradition of the New*, New York, Horizon, 1960.

Ross, A., *The Chicago Gangster Theory of Life, Nature's Debt to Society*, London, Verso, 1994.

Rowe, D., *Popular Cultures*, London: Sage, 1995.

Roweis, S. T., *Knowledge-Power and Professional Practice*, in Knox (ed.), 1988.

Rüedi, K. *et al.* (eds), *Desiring Practices, Architecture, Gender and the Interdisciplinary*, London, Black Dog, 1996.

Rueschemeyer, D., *Power and the Division of Labour*, Cambridge, Polity Press, 1986.

Rutter, C., *The English Cottage*, London, Batsford, 1926.

Salaman, G., 'Architects and their Work', *Architects' Journal*, Vol.21, No.1, 1970.

Samuels, I., 'The Plan d'Occupation des Sols for Asnières sur Oise: A Morphological Guide', in Hayward and McGlynn (eds), 1993.

Sanoff, H. and Cohn, S. (eds), *Proceedings of the First Annual EDRA Conference*, Raleigh, North Carolina State University, 1969.

Schelling, F. von, *Philosophie der Kunst*, discussed in Schröter, M. (ed.), *Werke*, 3rd supplementary volume, Munich, 1959, 223 ff.

Schröter, M. (ed.) *Werke*, 3rd supplementary volume, Munich, 1959.

Sennett, R., *The Uses of Disorder: Personal Identity and City Life*, Harmondsworth, Penguin, 1973.

Sitte, G., *City Planning According to Artistic Principles*, London, Phaidon, 1965.

Slim, H. and Thompson, P., *Listening for a Change: Oral Testimony and Development*, London, Panos, 1993.

Sola Morales, M. de, 'Another Modern Tradition: From the Break of 1930 to the Modern Urban Project', *Lotus International*, No.64, 1989.

Southwood, T. R. E., 'The Number of Species of Insects Associated with Various Trees', in *Journal of Animal Ecology*, Vol.30, 1961.

SPARC, *Portals, Bridges and Gateways: Cultural Explainers*, Los Angeles: Social and Public Art Resource Centre, 1994.

Stamps, A. E., 'Jungian Epistemological Balance', *Journal of Architectural Education*, 48/2, 1994.

Storey, J. (ed.), *Cultural Theory and Popular Culture*, New York, Harvester Wheatsheaf, 1994.

Stravinsky, I., *Poetics of Music in the Form of Six Lessons*, Cambridge, Mass., Harvard University Press, 1970.

Stübben, J., 'Der Städtebau', in *Handbuch der Architektur*, Darmstadt, 1890.

Stull, W. J., 'Community Environment, Zonkng and the Market Value of Single-Family Homes', *Journal of Law and Economics*, Vol. 18, pp. 535–57, 1975.

Tarn, J. N., *Five Percent Philanthropy: An Account of Housing in Urban Areas Between 1840 and 1914*, Cambridge, Cambridge University Press, 1973.

Taverne, E., 'Towards an Open Aesthetic, Ambities in de Nederlandse Architectuur 1948–1959', in Leupen, B. *et al.* (eds), 1990.

Teymur, N. *et al.* (eds), *Rehumanising Housing*, London, Butterworth, 1988.

Therborn, G., *The Ideology of Power and the Power of Ideology*, London, Verso, 1980.

Thrift, N., *Spatial Formations*, London, Sage, 1996.

Times, 25 January 1995.

Times, 15 October 1995.

Timms, E. and Collier, P. 'Treason of the Intellectuals?, Benda, Benn and Brecht', in Timms and Collier (eds), 1988.

Timms, E. and Collier, P. (eds), *Visions and Blueprints, Avant-Garde Culture and Radical Politics in Twentieth-Century Europe*, Manchester, Manchester University Press, 1988.

Townsend, S., *The Growing Pains of Adrian Mole*, London, Methuen, 1984, 29.

Tschumi, B., *Architecture and Disjunction*, Cambridge, Mass., MIT Press, 1996.

Tsoskonoglou, E., *Spatial Vulnerability to Crime in the Design of Housing*, unpublished PhD thesis, London, UCL, 1995.

Turner, L., 'Masonry and Stone-Carving', *Architectural Review*, Vol.59, April 1926.

Vasak, L., *Architectural Reinforcement of City Image*, unpublished MA dissertation, Oxford Brookes University, Joint Centre for Urban Design, 1989.

Venturi, R., *Iconography and Electronics Upon a Generic Architecture, A View from the Drafting Room*, Cambridge, Mass., MIT, 1996.

Vidler, A., *The Writing of the Walls: Architectural Theory in the Late Enlightenment*, Princeton, Princeton Architectural Press, 1989.

Washbrook, I., *Territoriality, Security and Public Space*, unpublished MA seminar paper, Oxford, Joint Centre for Urban Design, 1994.

Watkin, D., *Morality and Architecture*, Chicago, University of Chicago Press, 1977.

Weber, R., *On the Aesthetics of Architecture*, Aldershot, Avebury, 1995.

Weissman, G., 'Philip Morris and the Arts, Remarks by George Weissman', in *The First Annual Symposium: Mayor's Commission on the Arts and Business Committee for the Arts*, Denver, 5 September 1980.

Weston, R., *Modernism*, London, Phaidon, 1996.

White, C. L., *Women's Magazines 1693–1968*, London, Joseph, 1970.

Whyte, W. H., *The Social Logic of Small Urban Spaces*, Washington, DC, The Conservation Foundation, 1980.

Wilson, C. St J., *The Other Tradition of Modern Architecture, The Uncompleted Project*, London, Academy Editions, 1995.

Wolfe, T., *From Bauhaus to Our House*, London, Picador, 1992.

Wright, F.L., *Genius and the Mobocracy*, New York, Duell, Sloane and Pearce, 1949.

Wright, G., *The Politics of Design in French Colonial Urbanism*, Chicago, University of Chicago Press, 1991.

Wright, P., *On Living in an Old Country*, London, Verso, 1985.

Wynne, B., 'May the Sheep Safely Graze?', in Lash, Szerszysnki and Wynne (eds), 1996.

Young, C. in Gottdiener and Pickvance (eds), 1991, 202.

Zola, E., *Au Bonheur des Dames*, Paris, 1883, cited in Pevsner, 1979, 269.

Index